Relativity without Spacetime

Joseph K. Cosgrove

Relativity without Spacetime

palgrave
macmillan

Joseph K. Cosgrove
Philosophy
Providence College
Providence, RI, USA

ISBN 978-3-030-10240-1 ISBN 978-3-319-72631-1 (eBook)
https://doi.org/10.1007/978-3-319-72631-1

Cover credit: Tetra Images / Getty Images

This Palgrave Macmillan imprint is published by the registered company Springer Nature Switzerland AG
The registered company address is: Gewerbestrasse 11, 6330 Cham, Switzerland

For Natalia, a thinker

The original version of this book was revised. An erratum to this book can be found at https://doi.org/10.1007/978-3-319-72631-1_9.

Contents

Note on Citations

I have given citations in endnotes in the "author-date" style. Where I have consulted translations or other versions of a specific work, I list in square brackets, following the date of publication, the original date of the edition used. For example, I cite the Cohen and Koyré translation of the 1726 third edition of Newton's *Principia* as "Newton 1999 [1726]." Similarly, the standard Perrett and Jeffery translation of Einstein's 1916 review article on general relativity, from *The Principle of Relativity*, I cite as "Einstein 1952 [1916]."

List of Figures

CHAPTER 1

Introduction: A Critique of Minkowski Spacetime

In 1908, three years after Einstein first published his special theory of relativity, the mathematician Hermann Minkowski introduced his four-dimensional "spacetime" interpretation of the theory. Einstein initially dismissed Minkowski's theory as a "piece of mathematical trickery," remarking wryly that "[s]ince the mathematicians have invaded the theory of relativity I do not understand it myself anymore."[1] Yet Minkowski's theory soon found wide acceptance among physicists, including eventually Einstein himself—his conversion to Minkowski engendered principally by the realization that he could profitably employ it for the formulation of his theory of gravity (the general theory of relativity). Thus in his popular book on the theory of relativity published in 1916, Einstein famously remarks that general relativity "would perhaps have gotten no farther than its long clothes" if it had not been for Minkowski's innovation.[2]

The physical validity of Minkowski's concept of merged spacetime has rarely been questioned by physicists or philosophers since Einstein incorporated it into his theory of gravity. This is strange in one sense, for there is no general agreement on the physical meaning of Minkowski's theory. Indeed, one can only sympathize with Vesselin Petkov's complaint that while physicists and philosophers of science routinely endorse and employ Minkowski's four-dimensional formal apparatus, at the same time they habitually speak about the world as if it were really three-dimensional in the usual sense, giving back to themselves with one hand, as it were, what they have just taken away with the other.[3] Physicists in fact often employ

J. K. Cosgrove, *Relativity without Spacetime*,
https://doi.org/10.1007/978-3-319-72631-1_1

Minkowski spacetime with little regard to the whether it provides a true account of the physical world as opposed to a useful mathematical tool in the theory of relativity, while philosophers, for their part, sometimes treat the philosophy of space and time as if it were merely an appendix to Minkowski's theory. The purpose of this book, then, is to subject the concept of spacetime to a much overdue critical examination, with a view toward a more physically intelligible interpretation of Einstein's special and general theories of relativity. For I believe that Einstein's initial assessment of Minkowski was essentially correct.

Anyone conversant with the theory of relativity knows, or at least thinks they know, that Minkowski "merged" space and time into a single four-dimensional geometrical continuum. However, the precise character of the intended merging of space and time is not always clear in the literature on spacetime. One often reads, for example, that from our present vantage point we can see that even Newton and Galileo employed a "four-dimensional spacetime continuum," even if they treated time and space independently of one another. Einstein himself, for instance, suggests that the idea of a four-dimensional continuum is not something newly introduced by the theory of relativity, since classical mechanics also employed a four-dimensional continuum which, however, "falls naturally into a three-dimensional and one-dimensional (time), so that the four-dimensional point of view does not force itself upon one as *necessary*."[4]

It would be helpful at the outset, then, to note briefly some of the senses in which it may be said that space and time are "unified" in a given theory of physics. The minimal notion of such unification is a mere "n-dimensional manifold" in which each point event can be associated with *n* numbers (coordinates). This amounts to saying that every event happens at a time and place. In this minimal sense, Newton's physics indeed may be said to employ a four-dimensional spacetime manifold, something naturally suggested by diagrams in which time is symbolically represented by the length of a line in space (usually in Cartesian coordinates). But there is no single *continuum* of space and time in Newtonian physics, since space and time intervals are independent of one another, with no four-dimensional interval between events defined and therefore no metrical unification effected. While it is commonplace today to reconstruct Newtonian physics using the mathematical apparatus of differential geometry, as Michael Friedman notes, "We effect a relativistic unification of space and time only if we view space-time as a four-dimensional *semi-Riemannian* manifold."[5]

Einstein's 1905 special theory of relativity introduces a new kind of *metrical entanglement* of space and time. Here the measure of a time interval between two events in some inertial frame *A* depends, through the Lorentz transformation, on *both* the time and distance between those events as measured in an inertial frame *B* in motion relative to *A*, and likewise for a distance in *A* as a function of time and distance measured in *B*. As Einstein formulates it, the "necessity" of the four-dimensional point of view in special relativity, unlike the situation in Newtonian dynamics, lies in the "formal dependence between the way the space coordinates, on the one hand, and the temporal coordinates, on the other, have to enter into the natural laws."[6] By contrast, in pre-relativity mechanics, metrical intervals of time and distance are the same for all inertial reference frames.

Nevertheless, Einstein 1905 special relativity in no sense merges space and time into a *single continuum*. That is to say, even though in pre-Minkowski special relativity we regard the metrical properties of space and time as interdependent or entangled, space and time themselves remain *distinct* continua with no metrical unification per se. Thus, while in differential geometry, for instance, a metric continuum is defined by its distance function or line element (quadratic differential form), in Einstein 1905 there is no such distance function for time and space taken together. This point is easily obscured by the standard use in special relativity of Minkowski diagrams (which would be much more aptly termed "special relativity diagrams," since they in no wise distinguish Minkowski's theory conceptually from Einstein 1905), and the associated jargon of "world lines," "light cone structure," and so forth.

At the very least, then, it seems the term "spacetime" is associated with a number of distinct conceptions of the unification of space and time. But it is only with Minkowski's introduction of the invariant "interval" or four-dimensional displacement vector that we encounter anything that could be properly termed the unification of space and time into a *single continuum*. By "spacetime" in this study, then, I shall always intend specifically "Minkowski spacetime" or the idea, set forth most famously by Minkowski in his Cologne lecture of 1908, that space and time *as they physically exist* are merged into a single continuum, geometrically determined by a four-dimensional line element analogous to the Pythagorean line element of standard differential geometry.[7] To be sure, today we often speak of spacetime in a wider sense. However, the present terminological restriction is not at all arbitrary, for only with the advent of Minkowski's theory do we find the essential condition for a metric continuum of space and time: a distance function. Clearly there is no such notion in Galileo or Newton, and therefore it is misleading to say that either of them employed a "four-dimensional continuum."

It is unfortunate, in view of the very radicalness of Minkowski's proposal, that Einstein's 1905 special theory of relativity has come to be regarded as virtually synonymous with Minkowski's 1908 theory, as if the latter simply elaborated what was already implicit in the former. Such an identification, which forecloses the possibility that one might wish to avail oneself of Einstein 1905 while abstaining from Minkowski 1908, reflects a kind of whiggish view of the history of relativity, according to which Minkowski's theory represents the inevitable disclosure of the deep geometrical structure of Einstein's 1905 theory, originally overlooked by Einstein himself. But the proposition is a dubious one. For even if Minkowski's theory yields the same empirical results as Einstein 1905, the former nevertheless posits a new set of absolute geometrical objects that have no role whatsoever in the ontology of Einstein's original theory—the absolutes of which are rather the laws of nature. Thus, unless we wish to insist that theories making the same empirical predictions are equivalent, regardless of their respective ontologies and conceptual structures, Einstein 1905 and Minkowski 1908 are clearly *different* theories.

The weightiest argument on behalf of the physical reality of Minkowski spacetime has always been Einstein's general relativity. For in general relativity the metrical properties of the gravitational field find expression, at least in the usual version, through a "semi-Riemannian" generalization of the Minkowski spacetime interval. It would thus appear that save for the concept of the Minkowski interval, Einstein's formulation of the gravitational field in terms "curved spacetime" could hardly begin. However, an additional category of evidence for Minkowski's theory is primarily philosophical. Minkowski spacetime is often regarded as resolving a set of philosophical paradoxes, regarding time in particular, engendered by Einstein's 1905 special theory of relativity. According to a popular textbook presentation, for example,

> [A]lmost all of the "paradoxes" associated with SR [special relativity] result from a stubborn persistence of the Newtonian notions of a unique time coordinate and the existence of "space at a single moment in time." By thinking in terms of spacetime rather than space and time together, these paradoxes tend to disappear.[8]

To be sure, it would redound greatly to the favor of Minkowski's theory if the concept of spacetime truly resolved such paradoxes.

However, we shall find on the contrary that the concept of Minkowski spacetime actually plays no vital role in Einstein's theory of gravity and can be of no utility whatsoever in resolving paradoxes of time in either special or general relativity. Nevertheless, as Julian Barbour observes, "Minkowski's ideas have penetrated deep into the psyche of modern physicists ... [who] find it hard to contemplate any alternative to his grand vision."[9] Minkowski's ideas arguably have concentrated the minds of philosophers even more. Indeed, Minkowski's merging of space and time into a single entity has been a veritable boon to philosophy, providing endless opportunity for metaphysical speculation about the true nature of reality beyond the naive deliverances of subjective human experience. Moreover, it is not just physicists' and philosophers' "psyches" that have been penetrated by Minkowski's ideas, but the very language and mathematical notation by means of which relativity theory is formulated. Thus one can hardly open a textbook on general relativity without running across the assertion that a tensor, for instance, is an "inner product of vectors" or a "mapping of vectors onto to real numbers," the theory of spacetime dealing exclusively with associated "geometrical objects." This even though a tensor was not regarded as a geometrical object by Ricci and Levi-Civita, creators of the absolute differential calculus, nor is the quadratic differential "line element" of general relativity actually derived from geometry (it is rather derived from the Lorentz transformation).[10]

Rarely has it been remarked in this connection that while Minkowski's theory is set forth in vector form, by evident contrast with Einstein's algebraic methods in the 1905 special relativity paper, Minkowski spacetime is, in fact, an essentially *algebraic entity.*[11] In this respect Minkowski spacetime is quite unlike its alleged analogue the Pythagorean Theorem, which can be but need not be represented algebraically; for Minkowski spacetime can only be represented algebraically. In general, there has been insufficient scholarly attention to the historical process by which modern symbolic algebra was assimilated into mathematical physics, a development which, against much opposition, spanned the second half of the seventeenth century and most of the eighteenth century. As the physicist and historian John Roche observes, in his valuable study *The Mathematics of Measurement* (1998), the clarification of concepts in mathematical physics requires just such a historically informed approach:

> Physicists, mathematicians or philosophers who study the foundations of a concept … with this end in view sometimes attempt its reconstruction with little investigation of its historical derivation. This, I believe, is difficult to carry through successfully given the many questions the concept was designed to answer, and the multiple layers, shifting meanings and elements of incoherence which a concept commonly accumulates in the course of its history. Most if not all of this can be invisible without historical excavation.[12]

The need for such "historical excavation," which some phenomenologists have aptly termed "desedimentation," may well find its very example in the concept of Minkowski spacetime. For up until now historical analyses of Minkowski's theory have in general taken for granted the symbolic-algebraic structure and representation of the concept. By contrast, Newton's concept of force, for example, both can be and in fact was represented non-algebraically by Newton himself, and has in this respect been subjected to thorough historical analysis. Central to the argument of this study, then, and developed primarily in Part II, is the thesis that the conceptual structure of Minkowski spacetime can be adequately understood solely in terms of mathematical developments in the sixteenth and seventeenth centuries, assimilated into mathematical physics in the seventeenth and eighteenth centuries, by means of which the Euclidean method of ratio and proportion traditionally regarded up to the time of Newton and beyond as the form of mathematical representation proper to the science of physics was transformed into algebraic equations or "formulas." The specific critique of Minkowski spacetime set forth in Part II thus serves also as a case study in symbolic-algebraic representation and its influence on modern science's conception of nature.

I am guided throughout this study by the example of Socrates, for whom the pertinent question is always the one suggested by what appears on the surface (*eidos*) of things. Indeed, much of what I have to say should strike the reader as more or less obvious and hardly worth pointing out. Nevertheless, it is just the surface of things that we most tend to overlook. Deep structures have their interest, to be sure, but if we miss the surface we will be unlikely to find our way through the depths.

Notes

1. Quoted in Sommerfeld 1970 [1949], 102. I have been unable to locate the source of Einstein's remark about "mathematical trickery." Perhaps it is apocryphal.
2. Einstein 1961 [1916], 63.
3. Petkov 2012, 3–4.
4. Einstein 1979 [1949], 55.
5. Friedman 1983, 34.
6. Einstein 1979 [1949], 55.
7. That is, the Minkowski "spacetime interval" or quadratic differential form $c^2dt^2 - dx^2 - dy^2 - dz^2$. For our purposes it will almost always suffice to give this expression in terms of just one spatial dimension x and omit the differential symbol d; thus $c^2t^2 - x^2$.
8. Sean Carroll, "Lecture Notes on General Relativity," 1997, accessed July 3, 2017, https://arxiv.org/pdf/gr-qc/9712019.pdf.
9. Barbour 1999, 138.
10. Ricci and Levi-Civita's original paper (1901) on the subject, entitled "Methods of the Absolute Differential Calculus and their Applications," regarded geometry as just one possible application of the calculus. A tensor ("system") was defined analytically in terms of the invariant transformation properties of quadratic differential forms. On this point see Norton 1992, Appendix, 302–310.
11. An exception is Martínez 2009, 383–384: "In Minkowski's interpretation, the concept of a vector summarized coordinate-analytic notions. Previously, vector theorists had advocated the priority of vectors by conceiving them as consisting *fundamentally* of direction and magnitude and only incidentally as expressible in terms of Cartesian coordinates."
12. Roche 1998, 5–6.

PART I

The Concept of Minkowski Spacetime

CHAPTER 2

Minkowski's "Space and Time"

Although by comparison with his two other published talks on the same subject Minkowski's Cologne lecture "Space and Time" remains at a less technical level of exposition, the Cologne lecture is entirely sufficient for our purposes and comes with the additional advantage of wider familiarity. Minkowski saw himself not merely as offering a more elegant mathematical formalism for what we now call the special theory of relativity, but more importantly as unveiling the deeper geometrical structure of that theory in terms of what he called the "absolute world." The principal question is the precise form of unification of space and time Minkowski achieves in the Cologne lecture, beyond what Einstein himself had already accomplished in 1905.

2.1 Minkowski and Göttingen Science

Historians of science have situated Minkowski's approach to relativity theory in the context of an early twentieth century movement toward formalism in physics.[1] For example, Leo Corry has shown that Minkowski's endeavors in relativity theory must be in understood from the perspective of David Hilbert's program for the axiomatization of physics. Scott Walter emphasizes disciplinary aspects of the movement, such as the rivalry in German universities in the latter decades of the nineteenth century between mathematics and the newly constituted field of "theoretical physics." The formal turn in early twentieth century physics is epitomized

J. K. Cosgrove, *Relativity without Spacetime*,
https://doi.org/10.1007/978-3-319-72631-1_2

by the aims and activities of the group surrounding Hilbert at Göttingen, prominent among which was Hermann Minkowski (1864–1909).

Hilbert's program was animated above all by the idea, most famously mentioned by Minkowski in the last sentence of his Cologne lecture, of a pre-established harmony between mathematics and physics, as well as the conviction that left to itself physics tends to progress in too haphazard a fashion and, consequently, should not be "left to physicists":

> … Hilbert believed that physicists tended to solve disagreements between existing theories and newly found facts of experience by adding new hypotheses, often without thoroughly examining whether such hypotheses accorded with the logical structure of the existing theories they were meant to improve. In many cases, he thought, this had led to problematic situations in science which could be corrected with the help of an axiomatic analysis of the kind he had masterfully performed for geometry.[2]

Such would be the service that a resurgent *mathematical* physics, a subdiscipline of mathematics exclusively concerned with the formal structure of physical theory, could render the discipline of theoretical physics. Or, in the words of Poincaré: "[T]o disclose to the physicist the concealed harmonies of things by furnishing him with a new point of view."[3]

The formal program of "Göttingen science" raises a number of issues for the interpretation of Minkowski's foray in relativistic physics.[4] For instance, is Hilbert's program essentially heuristic, with mathematics merely revealing the deep formal structure of physical theories and on that basis suggesting new avenues of investigation? Or are we to take the notion of "pre-established harmony" more literally, and more radically, as potentially effacing the very distinction between mathematics and physics such that one could meaningfully speak, for instance, of the reduction of physics to geometry? Einstein himself, even after he embraced Minkowski's theory, stopped short of endorsing such a reduction.[5]

In Minkowski's case, the dramatic opening of the Cologne address is often highlighted in favor of the latter interpretation ("Henceforth space by itself, and time by itself, are doomed to fade away into mere shadows …."[6]). However, the independent existence of space and time or their existence "by themselves" is already ruled out by Einstein 1905, with its metric "entanglement" of space and time. Damour suggests that the interpretation of Minkowski in terms of what we today call block time or "eternalism" is chiefly the work of others and not Minkowski himself, arguing there is no evidence that Minkowski believed spacetime rendered the flow of time itself an "illusory shadow."[7] Instead, according to Damour, Minkowski's importance lies in his realization of the revolutionary significance of a set of deep mathematical structures in special relativity, largely discovered already by Poincaré: "Without fully understanding what Einstein had done, nor (at least initially) what Poincaré had achieved … [Minkowski] was lucky to unearth elegant and deep mathematical structures that were implicitly contained in their (and others') work, and had the boldness to embrace with enthusiasm their revolutionary character."[8] This is less than a ringing endorsement of Minkowski's significance, though, since the revolutionary character of these formal structures in special relativity was already explicit in Einstein's 1905 theory, whether Minkowski realized it or not. Formal talk of "four-dimensionality" in itself does not constitute anything revolutionary about Minkowski's theory.

What is truly revolutionary, rather, is the concept of the four-dimensional displacement vector or "spacetime interval"—and the whole apparatus of four-vectors in spacetime erected upon it—taken not merely as a formal structure but as *physically real*. For otherwise the theory amounts to no more than a formal reworking of Einstein 1905, perhaps analytically more powerful and elegant than Einstein's formulation, but not of surpassing theoretical significance so far as physics itself is concerned. But if the Minkowski spacetime interval is a real quantity in nature, then space and time as we experience them must be illusions of some kind, and quite fundamental ones at that: for in that case there exist in nature itself no distances in space or intervals of time. In Minkowski spacetime, as Friedman stresses, "there are no such entities as temporal intervals (or spatial intervals)."[9]

Minkowski does appear to have understood his four-vectors as physically real, although sometimes it can be difficult to tell for sure, as when he fails to distinguish clearly between his four-dimensional apparatus itself and other specific results in physics he wants to claim neither Lorentz nor Einstein were able to achieve. (Minkowski erroneously asserts that Einstein "made no attack on the concept of space," for example.[10]) In what follows, though, I shall not be interested in the question of Minkowski's exact intentions. For the last century, Minkowski's theory, however interpreted in specifics, has been generally regarded as a true description of the physical world and, above all, as an essential component of the conceptual structure of Einstein's general theory of relativity. I shall challenge that consensus on both counts. To do so convincingly, of course, will require an explanation of why Minkowski's theory, if it does not provide a true account of the physical world, works so well in general relativity.

2.2 "Space and Time," Sections I and II

Minkowski launches the Cologne address with the suggestion that it might be possible by "purely mathematical" considerations to arrive at "changed ideas of space and time."[11] Noting that Newtonian mechanics exhibits a two-fold invariance, corresponding in the first place to changes in position of the spatial coordinate system and in the second to changes in its state of motion, Minkowski observes that these two groups of transformations (the familiar displacements and rotations of Euclidean geometry, on the one hand, and the so-called "Galilean transformations" between inertial frames on the other) have not previously been brought together into one. It is his intention to do just that.

The purely spatial transformation group yields the familiar invariant $x^2 + y^2 + z^2$ for any rectangular coordinate system. However, Minkowski remarks, the aforementioned Galilean transformation permits our assigning to the time axis of our coordinate system any direction we please, corresponding to the different inertial motions of the Galilean system of reference (77). How shall we combine these two transformations, the first with its requirement of orthogonality in space and the second with its directional freedom for the time axis? Minkowski continues, "To establish the connection, let us take a positive parameter c, and consider the graphical representation of $c^2t^2 - x^2 - y^2 - z^2 = 1$." This graphical representation (see Fig. 2.1 below) depicts the famous hyperbolic coordinate rotation, in which the rotation angle of the coordinate axes corresponds to the veloc-

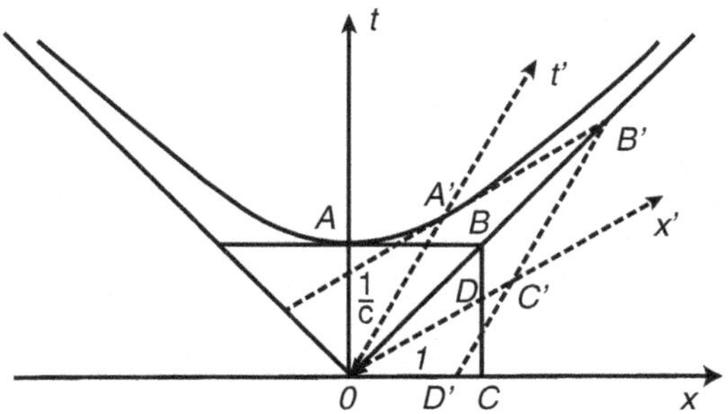

Fig. 2.1 Spacetime coordinate rotation

ity of some coordinate system in uniform motion relative to the original coordinate system.

For the spacetime coordinate rotation, the time axis t rotates clockwise to become oblique axis t' while the space axis x' rotates counter-clockwise to become oblique axis x'. The Lorentz equation $c^2t'^2 - x'^2 = c^2t^2 - x^2$ holds for any point on the hyperbola drawn through A, each such point representing the distance and time from the event represented at the origin of coordinates.

Of course, in itself the figure above is simply a graph of the Lorentz covariant equation $c^2t'^2 - x'^2 = c^2t^2 - x^2$; so let us concentrate rather on that equation itself, by which Minkowski proposes to treat time and space together geometrically by formal analogy to the invariant Pythagorean line element for space. Clearly, we cannot at the outset construe the parameter c in Minkowski's formula as the velocity of light. For such a procedure would *not* qualify as the derivation of new ideas about space and time by means of a "purely mathematical line of thought." Rather, Minkowski proposes to derive the Lorentz transformation in a purely mathematical way by means of an analogy with the Pythagorean Theorem.[12]

What initially would suggest itself as a "spacetime" analogue to the Pythagorean line element, though, would be $t^2 + x^2 + y^2 + z^2$. But this most obvious candidate presents an immediate impediment, namely, the heterogeneous dimensionality of the time and space variables. Clearly, the Pythagorean line element as a sum of squares is governed by the homogeneity requirement for addition as an arithmetical operation: quantities of unlike dimension do not add. Necessarily, then, if homogeneity is to be satisfied, Minkowski's "positive parameter c" must carry units of *velocity* (or distance per unit of time); for otherwise the algebraic expression for the interval would be arithmetically incoherent. We take careful note of the units of Minkowski's positive parameter c, for there exists a tendency in the literature to drop those units as if c were a dimensionless number. However, the units cannot be dropped if the expression is to remain physically intelligible. Thus adjusted, and with the signs of the spatial variables reversed (of which more below), the quantity $c^2t^2 - x^2 - y^2 - z^2$ is rendered invariant under what is called a "hyperbolic rotation" of the spacetime coordinate system, just as the Pythagorean $x^2 + y^2 + z^2$ is invariant under the familiar Euclidean rotation.[13] (An obvious difference, of course, is that the hyperbolic rotation is merely formal and cannot actually be physically performed as can the Pythagorean rotation.) This very group of hyperbolic transformations Minkowski designates G_c.

The value of the parameter c remains to be determined. Were c infinitely large the group G_c would reduce to the Galilean group. Much more intelligible, Minkowski suggests, is a finite and determinate, albeit very large, value for c (79).[14] Therefore, Minkowski concludes,

> [T]he thought might have struck some mathematician, fancy-free, that after all, as a matter of fact, natural phenomena do not possess an invariance with the group G_∞, but rather with the group G_c, c being finite and determinate, but in ordinary units of measure, *extremely great*. Such a premonition would have been an extraordinary triumph for pure mathematics. Well, mathematics, though it now can display only staircase-wit, has the satisfaction of being wise after the event, and is able, thanks to its happy antecedents, with its senses sharpened by an unhampered outlook to far horizons, to grasp forthwith the far-reaching consequences of such a metamorphosis of our concept of nature. (79)

Thus, with no prior inkling of the empirically determined velocity of light, the mathematician might well have derived the Lorentz transformation and with it the special theory of relativity itself.

Minkowski's preceding line of argument is the first in a long succession of what could be called formal derivations of the Lorentz transformation. By a "formal derivation" I mean a derivation that, unlike Einstein 1905, does not rely on the empirical light postulate but deduces the Lorentz transformation by means of formal mathematical considerations. Introductory presentations of the theory of relativity as a rule go the other way, deriving the Lorentz transformation on the basis of the light postulate and the special principle of relativity; and then, with the "three-dimensional" version of special relativity in hand, they introduce the Minkowski four-dimensional spacetime formalism. We must also distinguish between the geometrical derivation proposed by Minkowski above and derivations of the Lorentz transformation based on the formal constraints of the principle of relativity, with no prior appeal to empirically contingent facts about light propagation. Such derivations, which have much to recommend them, I set aside here.[15]

Eddington (1923) presents a disarmingly simple, if utterly misleading, formal derivation of the Lorentz transformation starting with the generally covariant "fundamental quadratic form" of spacetime: $ds^2 = g_{11}dx_1^2 + g_{22}dx_2^2 + g_{33}dx_3^2 + g_{44}dx_4^2 + 2g_{12}dx_1dx_2 \; \ldots$[16] On the basis of this expression, we forthwith recover the spacetime line element for "flat" spacetime in a small region (where the g's can be regarded as constant): $ds^2 = dx_1^2 + dx_2^2 + dx_3^2 + dx_4^2$. Designating the $x_{1,2,3}$ coordinate differentials as spatial, Eddington observes, yields the standard Euclidean metric of space $ds^2 = dx^2 + dy^2 + dz^2$. The x_4 coordinate differential must then have something to do with time, although it may not actually be the coordinate time t. If our clock is at rest we shall have $ds^2 = dx_4^2$. Accordingly, dx_4 should be proportional to t, corresponding as it does to equal lapses of t on our rest clock. Therefore, "we express this proportion by writing $dx_4 = icdt$," where $i = \sqrt{-1}$ and c is "a constant" (14).[17] Our equation for the interval is now $ds^2 = dx^2 + dy^2 + dz^2 - c^2dt^2$. Finally, since experiment reveals a real velocity with the remarkable property of invariance in all inertial frames, we can regard our constant c as the velocity of light in a vacuum. The Lorentz transformation then follows in the usual way.

This is all quite breathtaking on Eddington's part. We wonder where the "fundamental quadratic form" came from in the first place, since if we do not yet possess even special relativity we can hardly *assume*, with respect to space and time taken together, a fundamental quadratic form imported from differential geometry. We also wonder how $\sqrt{-1}$ and the constant c found their way into the equation, given that in the original form all four variables were symmetrical. That is, what is it about time in this hypothetical "spacetime manifold" that it merits such special treatment a priori? Eddington remarks only, "Historically this transformation was first obtained for the particular case of electromagnetic equations," as if the original non-geometrical derivation is to be regarded as a mere historical contingency. In Einstein's 1905 theory, of course, the Lorentz transformation applies to all processes, not just electromagnetic ones. Moreover, it is only based on the Lorentz transformation already in hand, from Einstein's 1905 theory, that Eddington has been able to construct his "time-related" term *icdt*. For no other justification is as much as hinted at with respect to interpreting the "fundamental quadratic form" as a spacetime metric in the first place, much less for inserting *icdt* for the time-related variable. Thus, if Eddington has derived the Lorentz transformation from the general notion of a four-dimensional spacetime manifold, that derivation is surely a circular one.

Michael Friedman presents a more careful, but in the end I think no more convincing case for our deriving the Lorentz transformation from Minkowski spacetime instead of the other way around.[18] "The surest and clearest way to derive the Lorentz transformations," writes Friedman, is to

> start by postulating the geometrical structure of Minkowski space-time. We then look for the group of transformations of Minkowski space-time onto itself that will preserve the geometrical structure: this turns out to be the Lorentz group .... Finally, using the geometrical structure of Minkowski space-time, we define the class of inertial frames and show that any two inertial frames are related by a Lorentz transformation.[19]

It is no doubt true that if we initially "postulate" the structure of Minkowski space-time we have a clear and sure way to derive the Lorentz transformation. But on what basis would we postulate such structure if we do not yet possess the Lorentz transformation from Einstein's 1905 theory? Clearly, Friedman is dissatisfied with derivations of the Lorentz transformation based jointly on the special principle of relativity, the light postulate, and various and vague assumptions about the "homogeneity" of space and time.[20] However, if we initially postulate the geometrical structure of Minkowski spacetime, as Friedman recommends, we have *ipso facto* postulated the flatness of spacetime as well, and along with it the very assumptions he so frowns upon in standard derivations. In Friedman's account, that is, we have no more reason to initially "postulate" the geometrical structure of Minkowski spacetime than we did in Eddington's above.[21] The most one could say is that Minkowski spacetime might function as a "hypothesis" which, once experimentally confirmed, could be employed, according to the *ordo essendi*, to derive the Lorentz transformation. But a hypothesis is not a postulate, and apart from the prior possession of the Lorentz transformation it is utterly implausible to imagine anyone advancing, even as a hypothesis, the notion of an "interval in spacetime" described by the equation $ds^2 = c^2dt^2 - dx^2 - dy^2 - dz^2$. As Minkowski admits, here the mathematician can exhibit only "staircase wit." Instead one first notices, as did Minkowski, the formal analogy between the Pythagorean Theorem (when written down algebraically) and the *already in hand* Lorentz expression above. We thus conclude that there is no intelligible way to derive formally the Lorentz transformation from the structure of Minkowski spacetime that does not in some way presuppose the Lorentz transformation itself.

Nevertheless, it is fair to raise some questions as to the physical coherence of Minkowski's geometrical approach, even when taken merely as a hypothesis. To make a convincing case for G_c, without invoking Einstein's light postulate, the first challenge is to show that the geometrical approach to relativistic physics could represent anything beyond merely a formal-mathematical analogy between two algebraic expressions ($x^2 + y^2 + z^2$ and $c^2t^2 - x^2 - y^2 - z^2$). And here we straightaway encounter an awkward disanalogy between the two. Observe that the variables *x*, *y*, and *z* in the Pythagorean line element carry the same dimension (spatial length) as the line element *ds* itself. Otherwise, the line element could not resolve into components, since such resolution presupposes homogeneity of dimension between the quantity in question and its components. The analogy with the Pythagorean line element, accordingly, would suggest that the spacetime line element *ds* also should be dimensionally homogeneous with each of the variables *t*, *x*, *y*, and *z*, such that all carry the dimension of *spacetime*. In the event, however, not only do we lack for any intelligible conception of such a physical dimension or units as "spacetime," but the actual units of the so-called spacetime interval are in fact units of spatial length. After all, that very fact is the whole point of Minkowski's "positive parameter *c*," which renders the expression for the spacetime interval arithmetically coherent by transforming the time variable *t* into a spatial distance *ct*. But by introducing, through the addition of the constant *c*, the formula $c^2t^2 - x^2 - y^2 - z^2$, Minkowski is left with simply the difference between the square of the spatial distance between two events and the square of the distance that would be traversed at velocity *c* in the time interval between those two events. It is all space, not "spacetime." On the hypothesis that the spacetime interval is physically real, however, it should carry the dimension and units of spacetime, in which case each of the variables *t*, *x*, *y*, and *z* should carry the dimension and units of spacetime as well, which they do not.

In the end, if Minkowski proposes to derive the Lorentz transformation by purely formal or mathematical considerations he must adduce some rationale for believing that the expression $c^2t^2 - x^2 - y^2 - z^2$ can intelligibly be regarded as a physical quantity in its own right. But the very physical sense of the expression $c^2t^2 - x^2 - y^2 - z^2$ is wholly determined by the transformation equation from which it is originally derived. Independently of that transformation equation the expression carries no physical meaning whatsoever. Furthermore, while the transformation equation $x'^2 + y'^2 + z'^2 = x^2 + y^2 + z^2$ can indeed be applied when the Pythagorean Theorem is represented in Cartesian coordinates and written down algebraically, the Pythagorean Theorem itself concerns relations between geometrical squares, and makes no essential reference to any transformation equation at all. The Greeks themselves gave the theorem in "component-free" form, as it were, simply as a sum of geometrical squares. Minkowski's derivation, by contrast, seems infected by a kind of circularity. For in order for the derivation to make sense, we must already be convinced that $c^2t^2 - x^2 - y^2 - z^2$ is meaningfully construed as a physical quantity ("interval" in spacetime). But we have been given no reason to think so other than the derivation itself.

As has so often been observed, Minkowski's derivation of the Lorentz transformation in "Space and Time" is in keeping with the aims of Felix Klein's *Erlanger* program, with Minkowski in this case interpreting the set of "invariant objects" under the Lorentz group of transformations in terms of a "rotation" of the spacetime coordinates *x*, *y*, *z*, and *t*.[22] But what we are given in the first two sections of Minkowski's "Space and Time" is merely a geometrical *representation* of the Lorentz transformation. At the end of Section I, for instance, Minkowski notes that for a change in reference system we obtain a new instantaneous space, concluding on this basis that "three-dimensional geometry becomes a chapter in four-dimensional physics" (80). That is true enough, but that result is contained already in Einstein 1905 and implies no unification of space and time beyond Einstein's; certainly not any "merging" of space and time into a single continuum.[23] That is to say, the whole exercise thus far in Minkowski's paper involves no more than the *interdependence* or entanglement of space and time, inherent in the Lorentz transformation itself and physically interpreted by Einstein 1905, quite apart from a four-dimensional geometry.

2.3 "Space and Time," Section III

Minkowski's real innovation, which does go beyond Einstein 1905, comes at the outset of section III, where we meet with the assertion that "the world-postulate permits the identical treatment of the variables *x*, *y*, *z*, and *t*" (83). Minkowski forthwith introduces the notion of "four-vectors" in spacetime: "We now, on the analogy of vectors in space, call a directed length in the manifold of *x*, *y*, *z*, and *t* a vector …" (84). In accordance with this conception, and by formal analogy with the Pythagorean Theorem, the magnitude of the invariant "displacement vector" in spacetime will be given by the Lorentz invariant $\sqrt{c^2t^2 - x^2 - y^2 - z^2}$. To be more precise, the preceding expression applies specifically to "space-like vectors" for which the quadratic expression assumes a negative value, reducing to $\sqrt{-x^2 - y^2 - z^2}$ for a suitable reference frame in which events occur simultaneously. For "time-like" vectors, on the other hand, we employ $\frac{\sqrt{c^2t^2 - x^2 - y^2 - z^2}}{c}$, corresponding to the possible trajectory of a physical particle traveling at less than the speed of light. Thus, for a suitably chosen inertial frame in which a clock is at rest, the expression for the time-like vector reduces to $\frac{\sqrt{c^2t^2}}{c}$ or simply $d\tau$. Clearly, the otherwise anomalous divisor c appended to the time-like vector represents a concession to the desideratum of treating x, y, z, and t "identically" as vector components. Otherwise, the time-like component would be $c\tau$ (with units of length), conflicting with the claim that the world postulate enables "identical treatment of the four variables."

Further developing the notion of the time-like vector, Minkowski designates the integral $\int \frac{\sqrt{c^2dt^2 - dx^2 - dy^2 - dz^2}}{c}$ along the path of a particle the "proper time" (τ), which, as we know, is also the time read off a clock traveling with the particle (85). Therefore, the proper time interval between two events, registered by an inertial clock present at both those events, is proportional to $\sqrt{c^2t^2 - x^2 - y^2 - z^2}$. This fact, *nota bene*, has nothing to do with the expression $\sqrt{c^2t^2 - x^2 - y^2 - z^2}$ representing a spacetime interval or with proper time "measuring" such an interval. It is simply an algebraic manipulation of the Lorentz transformation. The concept of "proper time" thus has no relation per se to "four-vectors" or spacetime intervals, being simply the time interval registered on a clock, related to the time on some other clock (regarded as at rest) by $\tau = t\sqrt{1 - \frac{v^2}{c^2}}$. This result, once again, is already explicitly contained in special relativity 1905, where Einstein notes that if a clock moves in a closed curve with constant velocity, it will be retarded compared to an initially synchronous clock at rest.[24]

Minkowski concludes Section III of "Space and Time" with definitions of the four-velocity and four-acceleration as the first and second derivatives (with respect to proper time) respectively of the four-displacement, noting that these two vectors are "normal" to one another (that is, their scalar product is equal to zero) (85). It follows that the "four-force" is also normal to the four-velocity, which leads to some noteworthy results in Minkowski's treatment of special relativistic mechanics in Section IV.[25]

We would be remiss, however, if we proceeded to Section IV without pausing over the ambiguous status, noted above, of the time component of the four-displacement vector; for this ambiguity necessarily accrues to the rest of Minkowski's four-vectors. Employing the *t* convention for the time component of the four-displacement renders the time component $\frac{dt}{d\tau}$ of the four-velocity simply the dimensionless number $\frac{1}{\sqrt{1-\frac{v^2}{c^2}}}$, an unsettling prospect for a vector component in the physical sense (and a physical sense it must have in Minkowski's "absolute world"). By contrast, if we use the *ct* convention, writing for the time component of the four-velocity $\frac{d(ct)}{d\tau}$ rather than $\frac{dt}{d\tau}$, we are left wondering why the distance light *would travel* during an interval of proper time should have anything to do with a particular component of a body's four-dimensional velocity. One can always defend Minkowski's approach, of course, by saying that we should not take the term "velocity" so literally, that what we have here is simply with the rate of change of the time component with respect to proper time. However, if we are not really talking about "velocity" in a physical sense or at least something like it, why construct the quantity in the first place and call it the "four-velocity"? It is no wonder that one textbook admits the only reason for calling the "four-velocity" a velocity is that its *spatial components* are "closely related to the particle's ordinary velocity … which is called the three-velocity."[26]

A further problem with respect to finding a physically intelligible interpretation of the four-velocity and, *mutatis mutandi*, all the other four-vectors constructed upon the four-velocity, is the mixing of quantities measured in different reference frames. The proper time τ is, we must not forget, the time measured relative to the frame in which our clock is at rest, whereas the coordinate time *t* is measured relative to some other rest frame.[27] But what justification there could be for regarding as a physical quantity in its own right the rate of change of a coordinate quantity with respect to proper time? Rather, such unnatural structure in a vector component simply is forced upon Minkowski by the requirement that all admissible quantities be invariant "geometrical objects." The only way to obtain the requisite invariance is through hybrid quantities like the four-velocity.

It might seem attractive, and would certainly be in accord with accepted usage of the term vector, to regard Minkowski's four-dimensional vectors as strictly analytical entities or "vectors" in a mathematical sense alone—simply sets of numbers related by a transformation equation. Einstein himself so regards them in his Princeton lectures of 1921, noting that

> [t]he ensemble of three quantities, defined for every system of Cartesian coordinates, and which transforms as the components of an interval, is called a vector.... We can thus get at the meaning of the concept of a vector without referring to a geometrical representation.[28]

Clearly, however, the fact that a set of quantities *transforms* as an interval does not mean that it is an interval in the physical or geometrical sense. Moreover, with regard to the notion of "four-vectors" in spacetime it is worth recalling that a "vector" in the usual sense is a *single* directed quantity which, if we so desire, we may resolve into "components" in some coordinate system. A transformation law will then govern the derivation of the components in one frame as a function of the components in another. However, such a resolution into components is optional: the concept of a "vector" as a directed quantity does not depend on it. Moreover, given a set of quantities related by a transformation law, it is not necessarily the case that these quantities may be regarded as components of some single directed quantity (vector). The quantities appearing in a transformation equation, after all, are simply algebraic *variables* related one to another in a prescribed way.

Notwithstanding his pronouncements about the absolute world, then, Minkowski's four-vectors in spacetime give every appearance of being merely analytical or, if you will, "symbolic" vectors. That is, Minkowski's vectors are "directed" solely in the symbolic or configuration space he has constructed: the so-called direction (angle) of the four-velocity vector, for instance, is simply an index of the magnitude of a body's three-dimensional velocity as represented on a spacetime graph. Furthermore, in the four-dimensional vector calculus, no single directed four-dimensional quantity in spacetime is identifiable such as could be subsequently resolved into components. Rather, Minkowski's four-vectors are and can only be constructed out of Cartesian components—or perhaps better, they exhibit "components" in a purely analytical sense but no directed resultant—and so they are coordinate-analytic in their very essence. From this perspective, it is surely ironic that Minkowski is often regarded as having departed from Einstein's algebraic approach in favor of a geometric or vectorial conception in mathematical physics. Minkowski employs the language of geometry, but his four-vectors are essentially analytical.

2.4 "Space and Time," Section IV

Up to the present point in Minkowski's presentation, all four-vectors have been constructed as ***kinematical*** entities, as it were, determined by formal-mathematical analogy with their three-dimensional counterparts. Thus, for the four-dimensional displacement vector, by analogy with the Pythagorean line element $\sqrt{dx^2+dy^2+dz^2}$, we obtain $\sqrt{c^2dt^2-dx^2-dy^2-dz^2}$. Similarly, for the four-velocity vector we have, by analogy with $\frac{dx}{dt},\frac{dy}{dt},\frac{dz}{dt}$, the four-velocity $\frac{dx}{d\tau},\frac{dy}{d\tau},\frac{dz}{d\tau},\frac{dt}{d\tau}$; and similarly for the four-acceleration. Minkowski devotes the remaining two sections (IV and V) of "Space and Time" to establishing that the four-vector approach can be consistently applied beyond kinematics to mechanics (Section IV) and electrodynamics (Section V); and furthermore, that his theory supplies a superior theoretical understanding in both these branches. For our purpose, since special relativity has no more essential connection to electrodynamics than to mechanics or any other branch of physics, and since the proposed explanatory superiority of Minkowski's theory applies equally to both or neither, let us, with a view toward our eventual discussion of the energy tensor in general relativity, concentrate on Minkowski's treatment of mechanics in Section IV.

We may, on the preceding pattern of defining four-dimensional vectors by analogy with their three-dimensional counterparts, write $\boldsymbol{f} = m_0\boldsymbol{a}$ for the four-force (m_0 designating the rest mass or "proper mass"), yielding what Minkowski terms the "force vector of motion" (*Kraftvektor der Bewegung*) (87). The latter may be regarded as the four-dimensional "effect" of the applied force or "motive force vector" (*bewegender Kraftvektor*). Minkowski now introduces his four-dimensional law of motion $f_{motive} = f_{motion}$. The suggested analogy with Newton's law of motion is inescapable. Newton writes, "A change of motion is proportional to the motive force [*vi motrici*] impressed,"[29] (Newton's "change of motion" corresponding to our change of ***momentum***), such that our version of Newton's law likewise can be written $F_{motive} = F_{motion}$ or $F_{\text{motive}} = \frac{dp}{dt}$. Clearly,

Newton's law of motion relates "motive force" as *cause* to change of motion as *effect* (Definition 4: "Impressed force [*vis impressa*] is the action exerted upon a body to change its state ..."[30]). In Newtonian terms, then, given the definition of force as an action on a body that changes its state of motion, we have two ways of *quantifying* force or saying how much of it there is in a given instance.[31] With respect to the acting cause, we select some particular force such as gravity and then write $F_g = \frac{Gm_1m_2}{r^2}$ (or, in non-algebraic terms such as Newton employed himself: the ratio of gravitational forces is proportional to the ratio of masses and inversely proportional to the duplicate ratio of the distance between those masses[32]). Likewise, with respect to the effect we can quantity force in terms of the magnitude of the effect produced, thus $f = ma$ (or "the force is as the mass and the acceleration conjointly"). Based on the proposed analogy with Newton's law of force, then, we would expect Minkowski to offer an intelligible account of the four-dimensional "motive force vector" as an *acting cause*, comparable to Newton's motive force (or "impressed" force).

Minkowski commences his discussion of the four-dimensional law of motion by asking, "When a force with the components X, Y, Z parallel to the axes of space *acts* [my italics] at a world-point P (x, y, z, t) ... what must we take this force to be when the system of reference is in any way changed?" (86). I take Minkowski here to be speaking of the "relativistic force," which is clear for a couple of reasons: In the first place, since only spatial components are listed, it cannot be the four-force. Moreover, the group G_c is explicitly assumed and so Minkowski cannot be referring to Newtonian force.[33] Since the relativistic force is not invariant, though, the four-force must be obtained by multiplying the components of the relativistic force by $\frac{dt}{d\tau}$ (the relativistic factor γ or $\frac{1}{\sqrt{1-\frac{v^2}{c^2}}}$), yielding an invariant vector γX, γY, γZ, γT for the motive force vector. Then for the "force vector of motion" the analogy with the right-hand side of Newton's law yields $f_{\text{motion}} = dp/d\tau$. Finally, we obtain for Minkowski's law of motion $F\gamma = dp_i/d\tau$ or, in terms of rest mass, $f_{\text{motion}} = \frac{d(m_0 v)}{dt}\gamma^2$. Clearly, Minkowski is seeking more than merely a relativistic transformation law for the force vector components X, Υ, Z, in which case standard Lorentz covariance would be sufficient.

The genuine beauty of Minkowski's four-vector calculus begins to emerge fully when we more closely examine the time component of the motive four-force γT. Minkowski continues:

> When the system of reference is changed, the force in question transforms into a force in the new space coordinates in such a way that the appropriate vector with the components $\gamma X, \gamma \Upsilon, \gamma Z, \gamma T$, where $T = \frac{1}{c^2}\left(v_x X + v_y Y + v_z Z\right)$ is the rate at which work is done by the force at the world-point P divided by c^2, remains unchanged.[34]

(Minkowski obtains T in the passage above by setting to zero the scalar product of the four-force and the four-velocity, such that $f_t = \frac{Fv\gamma}{c^2}$; and since $f_t = T\gamma$, we obtain $T = \frac{Fv}{c^2}$.[35])

Adepts in relativity will recognize Minkowski's introduction of what we know today as the "energy-momentum four-vector." Yet Minkowski never says that the time component of the four-momentum *is* energy, but that the time component divided by c^2 is energy. Since $Fv = \frac{dE_k}{dt}$ (E_k designating relativistic kinetic energy), it follows that the time component of the motive four-force is equal to $\frac{1}{c^2}\left(\frac{dE_k}{dt}\right)$.[36] If we then multiply the time component of the four-momentum by c^2, we obtain the relativistic kinetic energy or $mc^2\gamma$, which yields the approximation $mc^2 + \frac{mv^2}{2}$ for velocities small in comparison to c. Minkowski thus calls his fourth equation ($\frac{1}{c^2}mc^2\gamma = ma_t$) the "law of energy," by which he presumably means that just as the Newtonian second law of motion gives the rate of change of momentum as a function of the motive force, the time component of his four-dimensional law of motion gives the rate of change of energy as a function of the motive force vector. Moreover, since the four-acceleration vector is normal to the four-velocity, the fourth equation can be regarded simply as a consequence of the other three. Accordingly, Minkowski observes, the fourth equation implicitly contains "the whole system of the equations of motion" (87). Indeed, this point is more easily appreciated without any reference to operations on four-vectors: Since $\gamma = \frac{1}{\sqrt{1-\frac{v^2}{c^2}}} = f(v)$, it follows that $m\frac{d\gamma}{d\tau} = m\frac{d\left[f(v)\right]}{d\tau}$. And since the spatial part of the four-force is $m\frac{dv}{d\tau}$, the time component is determined entirely by the spatial components. One could therefore be forgiven for wondering whether the real lesson of Minkowski's cogitations on the four-dimensional law of motion is rather that the time component of the four-momentum (or any other four-vector) is simply redundant, with no additional insight into relativistic mechanics thereby achieved.

It is difficult to ignore the carefree manner in which expositors of the four-momentum, unlike Minkowski himself, toss around the constant c; and not simply as a function of the convention chosen for the time component of the four-displacement (that is, t or ct). This is all the more remarkable if we recall that it was only the "positive parameter c," with its units of velocity, that got the whole four-dimensional apparatus off the ground, as it were, in Section I of "Space and Time." Obviously, the constant c affects the physical meaning of any algebraic expression in which it appears, and if the time component of the four-momentum is to be regarded as energy (as per the expansion above for $mc^2\gamma$), then a factor of either c or c^2 must be simply thrown in to obtain the proper relativistic expression for energy; for the time component of the four-momentum itself, depending on your convention, is either $m\gamma$ or $mc\gamma$.

What physical justification is there for simply inserting c or c^2 in order to obtain energy? None that I can see. And the anomaly is in no wise obviated by our freedom to adopt the convention $c = 1$, even if some of the better authors seem to suggest just that. In his excellent introductory textbook on general relativity, for instance, James Hartle remarks that for a relativistic particle at rest, the energy $m\gamma$ expressed by the time component of the four-momentum "reduces to $E = mc^2$ in more usual units."[37] But this is simply untrue. If we are licensed to insert the factor c^2 merely based on having previously stipulated that $c = 1$, then we have just as much license to insert c^3 or c^4 or any other power of c we choose, obtaining for energy $E = mc^3$ or $E = mc^4$ and so forth as we please. In reality, using Hartle's convention $m\gamma$ for the time component of the four-momentum, the restoration of the "usual units" yields $mc\gamma$ for the time component of the four-momentum, not $mc^2\gamma$; and in that case the expansion for relativistic energy is $mc+\frac{mv^2}{2c}$, not the desired $mc^2+\frac{mv^2}{2}$. Similarly, Nerlich in his otherwise lucid account of Minkowski's theory informs the reader that for a particle at rest, the total energy, "when we restore the constant c^2 to remind ourselves of the conventional ratio between seconds and meters," is mc^2.[38] But the constant c^2 was never removed in the first place in Nerlich's own account, since Nerlich earlier gives the four-velocity as $\frac{ds}{d\tau}=c$, which means he is employing the *ct* convention. Once again, the *ct* convention yields *mc* for the time component of the four-momentum rather than mc^2.

Petkov, on the other hand, correctly writes $\frac{1}{c}\left(mc^2+\frac{mv^2}{2}\right)$ for the expansion of the four-momentum, such that on the *ct* convention the time component of the four-momentum is $\frac{E}{c}$.[39] Awkwardly, though, he then concludes that the *proportionality* of energy to the time component of the four-momentum reveals that the time component *is* energy. But the proportionality of two quantities in no wise implies they are forms of the "same thing." Force is proportional to acceleration in Newtonian mechanics, for instance, but that does not make force a form of acceleration (rather it is a *cause* of acceleration). After all, with $\frac{1}{c^7}$ as our constant of proportionality, mc^2 is proportional to mc^9 ($mc^2=\frac{1}{c^7}mc^9$), but surely we would hesitate to conclude on that basis that mc^9 is "really" a form of energy (or a form of any physical quantity at all).

It is, to be sure, merely a convention whether light velocity *c* enters our equations in special relativity with units of "meters per second" or "miles per hour" or anything else you please. But just as clearly, the fact that *c* enters our equations with units of *velocity* or "distance per unit of time" is not a convention. No physical velocity is merely a dimensionless number, after all, and, as we just recalled, Minkowski's geometrical approach is rendered feasible in the first place solely through his introduction of the "positive parameter *c*" *in units of space and time*; for otherwise the whole theory would be incoherent on purely arithmetical grounds. Consequently, to regard *c* as a "dimensionless constant" is something supporters of

Minkowski spacetime are least of all in a position to do.[40] In truth, the necessity of restoring *c* when it has been dropped has nothing to do with "reminding" ourselves of the convention we chose for the units of *c* (be those units meters, yards, seconds, light years, or anything else). As long as *t* stands for time and *x* stands for distance, there is no physical sense, in any units, to the algebraic expression $t^2 - x^2$. That is why the units of *c* must be restored, even if we require no reminder of the convention we have adopted because we have adopted the simplest one: $c = \frac{1\,\text{unit distance}}{1\,\text{unit time}}$. And, even were we for some physically inexplicable reason (aside from notational convenience), to drop the units of light velocity and treat the velocity of light itself as a dimensionless constant, there still is nothing special about the number 1. We could with as much justification stipulate that the velocity of light is the "dimensionless number" 300,000 or 186,000 or 3.14 or any other dimensionless number we choose. Surely it would be extraordinary if special relativity, which so famously eschews privileged inertial reference frames or coordinates, singled out a privileged system of units with $c = 1$!

There is no denying the elegance of Minkowski's derivation of the "law of energy" from the time component of the four-momentum. But has he succeeded in providing a four-dimensional analogue to Newton's second law of motion? Or has he simply summarized with a strikingly new formalism what is already contained in the 1905 special relativistic version of the relativistic force law $F = ma\gamma$? It appears to me to be the latter. Newton defines force as "an action on a body to change its state of motion," and even if we do not know the nature of the force, as is the case for Newton in his theory of gravity, the definition is still intelligible in terms of the concept of causality. No mere quantification of force, for instance with $\frac{Gm_1m_2}{r^2}$ as cause and *ma* as effect, can substitute for such a qualitative definition: for these very quantifications are physically intelligible only on the condition that we *already* possess a qualitative concept of force, however inadequate it might ultimately prove to be. By contrast, Minkowski supplies quantifications, but no physical concepts. For what could it mean for a four-dimensional force to "act" at a world-point? What concept of "action" has been defined for the absolute world in which "nothing moves" and thus nothing changes? It is not surprising, then, that Minkowski's four-dimensional "law of motion" is rarely if ever mentioned either in textbook accounts of special relativity or philosophical defenses of the Minkowski's theory. For a law of ***motion*** presupposes the reality of motion and in Minkowski spacetime, "[t]hese new four-dimensional objects do not move …."[41] By contrast, Einstein's 1905 special relativity maintains a physically intelligible concept of force, since $f = m_0 a\gamma$ holds in any inertial frame provided the acceleration remains parallel to the force. Seen in this light, it would appear that in the context of the special theory of relativity, at least, all Minkowski's mechanics accomplishes is to replace a coherent concept of force with an incoherent one. However, Minkowski's four-vector calculus does provide an invariant expression for energy and momentum taken together, neither of which is so formulated in Einstein's 1905 special relativity. Moreover, as we know, in his theory of gravity Einstein will avail himself of Minkowski's energy-momentum four-vector as the basis for the stress-energy tensor, just as he will employ the four-displacement vector as the basis for the metric tensor. These points we will have to address at the appropriate point in Chap. 7.

Notes

1. See for instance Pyenson 1977, Corry 1997 and 1998, and Walter 1999.
2. Corry 1997, 274–275. Such focus on logical consistency is indeed exemplified by Minkowski's address, as when, after introducing his "world postulate," Minkowski sets out to demonstrate that "the assumption of the group G_c for the laws of physics never leads to a contradiction …." (Minkowski 1952 [1909], 86).
3. Quoted in Pyenson 1977, 89.
4. On Göttingen science in general during the period in question, see Heelan 1987, sec. 2, 371–373.
5. See Lehmkuhl 2014.
6. Minkowski 1952 [1909], 75.
7. Damour 2008, 626–628.
8. Damour 2008, 629.
9. Friedman 1983, 306.
10. Minkowski 1952 [1909], 83.
11. Minkowski 1952 [1909], 75. Page references otherwise unidentified in this section are to Minkowski's "Space and Time" (Minkowski 1952 [1909]).
12. Various authors have noted Minkowski's failure in the Cologne lecture to mention Poincaré's contributions. Since it is Minkowski's formulation that was taken up into subsequent history of mathematical physics, for our pur-

poses we can regard Minkowski as the "inventor" of four-dimensional spacetime. On Poincaré's contributions to special relativity in relation to Minkowski see, for instance, Damour (2008), section II, and Walter (1999), section 2.2.

13. A hyperbolic rotation sweeps out angles on a hyperbola rather than on a circle as in regular trigonometry. In "Space and Time," Minkowski declines to use the hyperbolic functions *sinh* and *cosh*, although presumably he knew how to employ them. The hyperbolic functions leave the minus signs for the spatial variables intact and are in that respect less hospitable to the desired formal analogy with the Pythagorean Theorem.
14. Presumably a finite value is more intelligible because if c were infinitely large, then $\frac{dx}{dt}$ would reduce to $\frac{dx}{0}$. For an infinitely large velocity, that is, no time elapses during the traversal of any distance.
15. See, for instance, Mermin 1984.
16. Eddington 1965 [1923], 10.
17. Eddington actually writes for the time component dy_4^2, but for consistency I will adopt dx_4^2.
18. Friedman 1983, 138–142.
19. Friedman 1983, 138–139.
20. Friedman 1983, 142.
21. Friedman is not explicit that the structure of Minkowski spacetime is the only sure way to the Lorentz transformation, as opposed to simply the best way, although that is certainly the impression I get.
22. The prominence of Klein's program in Minkowski's approach to relativity is somewhat downplayed by Scott Walter (Walter 2014).
23. Einstein's definition of the length of a moving rod as the distance between the simultaneous coordinates of its endpoints essentially defines an instantaneous relative space.
24. Einstein 1952a [1905], 49–50.
25. The derivation can be found in any beginning textbook on relativity. Since the value of the four-velocity is c, it follows that $\boldsymbol{u}^2 = c^2$. If we then differentiate with respect to τ, we obtain using the chain rule $2(\boldsymbol{u} \bullet d\boldsymbol{u}/d\tau) = 0$ or $\boldsymbol{u} \bullet \boldsymbol{a} = 0$, and so $\boldsymbol{u} \bullet \boldsymbol{f} = 0$. As a rule, I shall identify four-vectors by boldface type.
26. Schutz 1985, 44.
27. Nerlich (2013, 188–122) argues that the proper time should be understood as a frame-independent quantity rather than as a quantity relative to any particular reference frame (or all of them). I shall address the distinction between frame *invariance* and frame *independence* in Chap. 3 below. Here I

simply observe that even on Nerlich's interpretation the four-velocity is a hybrid construction, since we take the derivative of a *relativistic* quantity (coordinate time) with respect to an "absolute" quantity (proper time).

28. Einstein 1953 [1922], 11.
29. Newton 1999 [1726], 416.
30. Newton 1999 [1726], 405.
31. Newton does not similarly distinguish between a qualitative definition of "quantity of motion" and the quantification of the same in a law of motion: "Quantity of motion is the measure of motion that arises from the velocity and the quantity of matter jointly" (Definition 2 of Newton 1999 [1726], 404). Presumably, the omission is because "quantity of motion" is already the quantification of the qualitative concept of motion. What would rather require qualitative definition would be *motion* itself, a concept Newton presumably takes as primitive and therefore indefinable.
32. Newton typically speaks of single quantities being "proportional" to other quantities, but obviously the proportionality of *ratios* is intended, since a proportion equates ratios, not quantities. Niccolò Guicciardini argues that Newton's practice of referring to "proportions" between single quantities at specific points (as for instance Prop. VI, Book I of the *Principia*) suggests that Newton actually derived these results by means of algebra. See Guicciardini 1999, 125–135. Today we express a proportion algebraically by means of a "constant of proportionality."
33. In the appendix ("Mechanics and the relativity postulate") to Minkowski's earlier and more technical paper on relativity (Minkowski 2012b [1908], 51–110) there is no such detour through relativistic force. One must presume that in "Space and Time" Minkowski is proceeding for the benefit of physicists unfamiliar with the four-vector calculus.
34. I have corrected the Perrett and Jeffery translation, where "divided by c^2" in the quoted passage incorrectly reads, "divided by c." The Lewertoff and Petkov translation (Minkowski 2012c) of "Space and Time" corrects this typographical error. I have also substituted γ for Minkowski's $\frac{dt}{d\tau}$ and vx for his $\dfrac{dx}{d\tau}\Big/\dfrac{dt}{d\tau}$ (and likewise for the other spatial variables).
35. $(cft, F\gamma) \bullet (c\gamma, v\gamma) = 0$, and so $c^2ft\gamma - Fv\gamma^2 = 0$ or $f_t = \frac{Fv\gamma}{c^2}$. Once again, F is the relativistic force.
36. This follows from the fact that $v_xX + v_yY + v_zZ = \frac{dE_k}{dt}$ by the work-energy theorem.

37. Hartle 2003, 87.
38. Nerlich 2013, 64.
39. Petkov 2005, 113.
40. Even Einstein himself, uncharacteristically, seems to have fallen prey to the temptation. In his *Autobiographical Notes*, for instance, Einstein remarks that if "one introduces as the unit of time, instead of the second, the time in which light travels 1 cm, *c* no longer occurs in the equations. In this sense one could say that the constant *c* is only an *apparent* universal constant. It is obvious and generally accepted that one could eliminate two more universal constants from physics by introducing, instead of the gram and the centimeter, properly chosen "natural" units (for example, mass and radius of the electron). If one considers this done, then only 'dimensionless' constants could occur in the basic equations of physics" (Einstein 1979 [1949], 59). This is all clearly false and represents a seduction by pure mathematics to which Einstein was not normally susceptible. An adjustment of units does not remove *c* from the equations in any sense save a *notational* one. Clearly, *c* still maintains its dimensionality even if adjusted to "one unit of distance per one unit of time." It is hard to imagine Einstein really thinking otherwise.
41. Nerlich 2013, 91.

CHAPTER 3

Special Relativity and Spacetime

The questions on Minkowski spacetime raised thus far can be framed in general terms as follows: What is the relation between the concept of Minkowski spacetime and its object—what it intends or is "about"? That relation is so far obscure, and in this chapter I shall elaborate that obscurity in three principal respects: first, the theory of spacetime articulates no clear meaning to the concept of a "single continuum" of space and time; second, the theory conflates two quite different types of geometrical representation—*graphs* and *images*—and so reads off visual features of graphs as if they were direct images of the physical world; and third, the theory of spacetime misconstrues the significance of *invariance* in special relativity by way a false analogy with geometry.

3.1 Spacetime and the Concept of a Continuum

From a more or less common sense perspective, we can define a continuum as an infinitely divisible magnitude of any kind (spatial, temporal, or otherwise). Slightly more formally, we might say that a *metric continuum* is defined by (1) a distance function between any two points and (2) the infinite proximity of neighboring points in the sense that for any given point, there exists a neighboring point such that the distance between the two is less than any given distance. In this general sense temperature, for instance, pressure, a line in space, loudness, time intervals, anything that admits of continuous degree, may be regarded as a continuum. A metric

J. K. Cosgrove, *Relativity without Spacetime*,
https://doi.org/10.1007/978-3-319-72631-1_3

continuum in differential geometry, in the usual sense, is a space with a line element or quadratic differential function giving the distance between any two points as a non-negative real number. In Riemannian geometry, this quadratic differential form is derived from the Pythagorean Theorem.

By its very nature, a metric continuum is a *single* continuum; for there can be no distance function between points of different continua. Were there such a function, *per impossibile*, its dimension or units would be undefined.[1] We do refer loosely to the "spacetime continuum" of Newton's theory, for instance, but with no four-dimensional metric Newtonian spacetime in reality comprises two distinct continua, one spatial and one temporal. The same applies to Einstein's special relativity 1905, in which space and time are metrically entangled but not unified into a single metric continuum.

A last criterion for a metric continuum is a transformation function from one measuring system to another. In the case of a one-dimensional continuum this is simply a matter of unit conversion and a reference point. To convert from Celsius to Fahrenheit in the temperature continuum, for instance, we multiply by 9/5 and add the constant 32; and so our transformation equation is $^{\circ}\mathrm{F} = \frac{9}{5}\,^{\circ}\mathrm{C} + 32$. For a geometrical continuum we require a transformation that accounts for displacements, rotations, and motions of the coordinate frames, expressing each transformed variable as a function of all the variables in the other system. Thus, in Cartesian coordinates, for a rotation in two-dimensional Euclidean space we obtain:

$$x' = x\cos\theta - y\sin\theta$$
$$y' = x\sin\theta + y\cos\theta$$

As David Bohm aptly observes,

> Without such a transformation, we would hardly even be justified in regarding the … dimensions as united into a single space as "continuum" (e.g., in an arbitrary graph, in which one physical quantity such as temperature is plotted against another, such as pressure, there is no such unification).[2]

Thus while I could indeed plot my beer consumption over time on a graph, I could not meaningfully speak of a "beer-time continuum."

In terms of the three criteria we have listed, Minkowski spacetime would appear to qualify as a (single) continuum. Its "points" are events, its distance function the Lorentz invariant $ds^2 = c^2dt^2 - dx^2 - dy^2 - dz^2$ ("spacetime interval"), and its transformation equations (in two dimensions and employing hyperbolic functions with $\gamma = \frac{1}{\sqrt{1-\frac{v^2}{c^2}}}$):

$$ct' = ct\cosh\gamma - x\sinh\gamma$$

$$x' = x\cosh\gamma - ct\sinh\gamma$$

Yet we immediately notice an awkward circumstance. The transformed quantities above are *ct* and *x*, not *t* and *x*. Otherwise those quantities would be of heterogeneous dimension, rendering senseless the arithmetical operation of subtraction. (That was the whole point of Minkowski's "positive parameter *c*.") We certainly expect all the dimensions of a single continuum to be of the *same kind*, in accord with the requirement for a distance function. And, indeed, it would appear that if continua of different dimension are to be "merged" into a single continuum, each must lose its specific dimensionality and assume the merged dimensionality. One way such a merging might be accomplished, formally at least, is by dropping the constant *c* and regarding the variables *t* and *x* as dimensionless numbers. Here a kind of "homogeneity of dimension" is accomplished by abstracting from the nature of the physical quantities under consideration. Short of that, a distance or interval, in the physical or geometrical sense, must carry some specific kind of dimension and be measured in a specific kind of units.

It follows that if space and time themselves, and not just the algebraic variables *t* and *x*, are truly to be merged into a single continuum, either (1) space must be "temporalized," (2) time must be "spatialized," or (3) both space and time must give way to a third type of merged dimension ("spacetime"). Moreover, as Émile Meyerson observed early on (1925) in the history of the theory of relativity, the fact that Minkowski's construction is a "geometry" means that the merging in question will necessarily be effected "to the advantage of space"; that is, option (2) above must prevail:

> It should be noted that if time and space are henceforth to be more or less merged into a single continuum, this change will clearly work to the advantage of space.... Let us observe, moreover, that this already follows from the fact that the construction at which one arrives is a *geometry*. And one need only open an exposition of the doctrine to note that, where time is concerned, the writer always speaks of one dimension, obviously conceived as spatial, while no attempt is ever made to represent the properly spatial dimensions in terms of time.[3]

In such a merged entity, time must acquire a geometrical or spatialized nature, which we nevertheless somehow experience as if events were "coming to be and passing away," all the time being ourselves confined to the present. Such a philosophy of time, often called eternalism or the "block view," is probably the dominant view today among philosophers of time.

Yet option (3) would arguably be the more logical choice, since, at least *prima facie*, the merging of time and space into a single continuum should favor neither time nor space. A physically real *spacetime interval*, that is, should carry the dimension of *spacetime* and accordingly be measured in units of spacetime, whatever those might be, rather than in units either of time or space. In that event, both space and time would be experiential illusions of some kind or at least "emergent phenomena" grounded in some more fundamental reality (itself neither spatial nor temporal.) Indeed, philosophical accounts of Minkowski's theory often seem to oscillate between option (2) and option (3). For obvious reasons, option (1), or the temporalizing of space, is not seriously entertained.[4]

The single continuum theory of space and time must also provide an intelligible account of the relation between four-dimensional spacetime and the three-dimensional world of Einstein 1905 special relativity. It is not enough to speak of a three-dimensional projection or "cross-section" of the four-dimensional absolute world which alone is truly real. In an actual geometrical projection, after all, a shadow for instance, the number of dimensions is reduced by one while the qualitative *type* of dimension (space) is left unaltered. In the alleged "spacetime projection," by contrast, the qualitative type of dimension (spacetime) is itself altered, such that an absolute, non-perspectival reality with "merged" dimensionality nevertheless exhibits, when so projected, distinct dimensions of space and time.

We must ask whether the theory of spacetime does not in reality conflate two quite distinct meanings of the "three-dimensional projection" of a four-dimensional absolute world: in one sense the literal geometrical resolution of a four-vector into its spatial and temporal components, and in a very different sense the *metaphorical* projection of the absolute world in *experience*, from the particular perspective of an "observer." When Minkowski asserts, for instance, that "only the four-dimensional world in space and time is given by phenomena, but the projection in space and in time may be undertaken with a certain degree of freedom," presumably he means that in choosing a specific inertial frame of reference we project the four-dimensional absolute world onto a three-plus-one-dimensional relative one.[5] This interpretation indeed is suggested by the "Minkowski diagram," with its depiction of multiple frames on the same graph.[6] The specific "relativity" at issue here clearly is a relativity with respect to the inertial frames licensed by the special theory of relativity—that is, the resolution of a spacetime four-vector into components relative to one of those frames—not a perceptual relativity with respect to some hypothetical "observer" who experiences the world as a three-dimensional present.

However, *ex hypothesi* in Minkowski's theory, only the four-dimensional world truly exists; and in that world there are solely spacetime intervals, not time intervals and spatial distances per se. How, then, can there obtain any such "projection" with respect to a three-dimensional inertial frame that according to the theory is not objectively real? The analogous projection of a three-dimensional object onto a two-dimensional surface (casting a shadow, for instance) can be freely undertaken only because two-dimensional surfaces *do exist* in three-dimensional space, and exhibit the same spatial quality of dimension as three-dimensional space. But here we have postulated a projection "onto" something that previously has been deemed not to exist except *as* a projection. We cannot give ourselves back with one hand what we have just taken away with the other. Thus, in the end, the "three-dimensional projection" of Minkowski's absolute world could only be a projection in subjective experience, for in Minkowskian special relativity there exists by hypothesis no three-dimensional physical world upon which such a projection could terminate. In that case, however, our ability to formulate special relativistic physics three-dimensionally with respect to inertial frames is rendered unintelligible.

3.2 The "Geometry of Spacetime": Graphs and Images

In the theory of Minkowski spacetime, much of the weight of the analogy between space and spacetime is borne by the so-called spacetime coordinate rotation. This graphic device is represented, once again, on the famous Minkowski diagram from the Cologne address (Fig. 3.1):

Fig. 3.1 Spacetime coordinate rotation

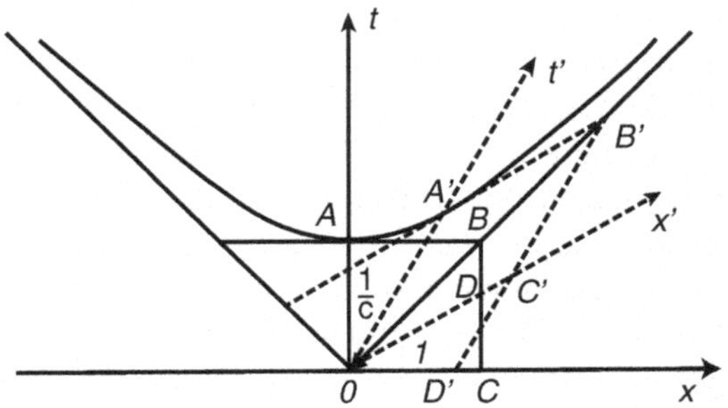

To review, Cartesian axes *x* and *y* are rotated "inward" to obtain oblique axes x' and y', with each point on the hyperbola drawn through *A* representing an event at a given time and place with reference to the origin. Just as for any point on a circle in regular Cartesian coordinates, the rotation of the coordinate axes leaves unchanged the quantity $\sqrt{x^2 + y^2}$ (the distance between the point and the origin), so for the hyperbolic rotation the quantity $\sqrt{c^2t^2 - x^2}$ (the so-called spacetime distance) remains invariant, regardless of the orientation of the coordinate axes.

To better appreciate the meaning of the "spacetime rotation," we must consider more closely the nature of geometrical representation in general. Let us distinguish, in admittedly oversimplified fashion but sufficient for present purposes, two basic types of geometrical representation: *images* and *graphs*. An image (or "icon") is a direct representation in the sense that it pictures what it represents, thus bearing a genuine resemblance to what it is about.[7] A circle drawn with pencil and compass, for instance, perhaps from the books of Euclid or Apollonius, qualifies as an image of a circle in space. The drawn figure or image is not the object of the proof, of course, since it is not strictly circular in a mathematical sense; but it does bear a direct resemblance to the geometrical circle in that it is "more or less" a plane figure with its points equidistant from the center.

Images have some interesting phenomenological features that have been analyzed by philosophers of art. A small child will point to a photo and say, "That's Daddy!" even though she very well knows that Daddy is not a flat piece of paper. She still discerns the genuine "sameness" between the image and Daddy. Thus an image exhibits a kind of identity in difference that renders present what is absent, by way of this sameness with the image. The relation is quite different from that between a mere sign and the thing signified. If I am driving down the interstate and see a sign that says "Boston 45," I now know that Boston is 45 miles away, but I do not point at the sign and say, "There's Boston," as I would if I picked out the city on a map of Massachusetts or saw a photo of the skyline. A mere sign "points" as it were, by convention, while an image *pictures* by virtue of its more direct relation to that which it represents.

The geometrical representation of something non-spatial can never be an image. However, still within the sphere of geometrical representation, and situated somewhere between images and "mere signs" with respect to the directness of the relationship with what it intends, lies the interesting category of *graphs*. An example is the "Bell Curve." The bell shape itself, maybe a line connecting plotted points designating tabulated data (for instance, frequencies of grades on an exam) does not in any way resemble the actual frequency of those grades. It is not an image of that frequency. But it evidently does more than merely point in the manner of a sign like the table of data itself. For in a way the Bell Curve still pictures something. Thus, one might point at the graph and say, "There's the distribution for last week's exam."

As compared with a genuine image, however, like a drawing of a circle, a graph represents indirectly or symbolically. We find an early example of a graph in the history of mathematical physics in Galileo's treatment of projectile motion (*Two New Sciences*, Theorem I, Proposition I of Day Four), in which on a single diagram Galileo represents time by means of a horizontal line and space by means of an intersecting vertical line. The parabolic shape of the plotted line is in this case only accidentally an image of the actual path *in space* of the falling body. Rather, Galileo's graph represents visually an idealized set of data (vertical positions of a projectile at points in time), or perhaps better an idealized law.[8] On the other hand, a plot of a parabolic trajectory in space, with both the horizontal and vertical axes representing distance in space, would be an image in the proper sense. And, indeed, the same diagram can be an image or a graph, or even both at once, depending on how we interpret the vertical and horizontal axes respectively. In Newton's *Principia*, for instance, we first encounter a graph in Lemma X of Section I, Book I, where times are represented on the vertical and velocities on the horizontal (with spaces traversed proportional to the areas of triangles). The very same diagram was an image in the previous Lemma IX, where Newton demonstrated a proposition about the geometrical relationship between the sides and areas of such triangles.[9] The advantage of a graph over simply tabulating data is that we can see patterns that might otherwise go unnoticed. The apex of the Bell Curve, for instance, indicates the most likely test score, and its shape calls our attention to how the different scores are smoothly distributed from lowest to highest.

Graphs such as employed in analytical geometry and mathematical physics are, as a rule, symbolic representations of *equations*, or at least they imply equations. Thus in the graph of the equation for a circle, $r^2 = x^2 + y^2$, the points on the graph represent relations between distances along the x and y axes, which themselves represent possible numerical values of the variables x and y, which in turn represent whatever we are actually talking about. The question therefore arises whether the drawn circle on the graph of the equation $r^2 = x^2 + y^2$ should be regarded as an image of a circle (direct representation) or a graph of a circle (indirect representation). In fact, the drawn circle can be taken, at different levels of representation, either as an image, a graph, or even both at the same time—although qua graph it is never an image per se. The equation itself defines general relations between numbers and does not necessarily have anything to do with a geometrical circle. For instance, were there a quantitative relationship in economics between, say, marginal tax rates and economic growth, such that the sum of squares of these quantities equaled a constant, economists no doubt would speak of the "circular relationship" between economic growth and tax rates. In that case our diagram would function purely as a graph of economic data. But if we are doing geometry, then the graph of the equation is simultaneously an image of the geometrical entity of interest—a circle in space. Geometry therefore comes into play here at two different levels of representation: as symbolic means of representation (graph of the equation) and as image of the thing being represented (a circle in space).

Descartes himself, originator of what today we call "graphing an equation," sets forth as clearly as one could want the symbolic conception governing what we nowadays call "analytical geometry":

> We have as much reason to abstract propositions from geometrical figures, if the problem has to do with these, as we have from any other subject matter. The only figures we need to reserve for this purpose are rectilinear and rectangular surfaces, or straight lines, which we also call figures, because, as we noted above, these are just as good as surfaces in assisting us to imagine an object that is really extended. Lastly, these same figures must serve to represent sometimes continuous magnitudes, sometimes a set or number.[10]

While we are used to thinking of analytical geometry as a method for doing geometry algebraically, which it is, to be sure, Descartes' decisive innovation is the symbolic representation of quantity in general by line lengths, which is why he begins his *Geometry* of 1637 with a demonstration of how to perform the basic arithmetical operations by means of geometrical figures.[11] So even when we really are doing geometry, we will "abstract" from geometrical figures by using lines (coordinate axes) to represent the general quantitative relationships determined by equations. Our figures or graphs will sometimes represent something geometrical, like a circle or ellipse, but sometimes not ("a set or a number"). The graph is therefore always first and foremost a *symbolic* space, regardless of whether ultimately it represents something geometrical.

Nevertheless, when we are doing geometry, a graph of the equation for a geometrical figure is always at the same time an image of that figure: for the geometrical representation of such an equation also pictures the figure. The situation is quite different when we are not doing geometry, though, for here graph and image never coincide. There can be no geometrical image of a non-geometrical object. Suppose I have a body in uniform motion, for example, satisfying $t = \frac{s}{v}$ (t = time, s = distance, v = velocity). In Cartesian coordinates, with time on the vertical axis and space on the horizontal axis, the graph of the equation for the body's trajectory is a straight line. But this straight line is not an image of the body's trajectory in space and time. For, after all, if the body were accelerating rectilinearly it would still be making a straight line in space, but its graph would no longer be straight line. Even to say that $t = \frac{s}{v}$ "graphs as a straight line" presupposes Cartesian coordinates. If we apply the coordinate transformation $s' = s^2$ the graph will now be a curved line, even though the body is still moving uniformly and making a straight line in the x direction.

Clearly, a Minkowski diagram is a graph rather than an image. It is never both at once, as in our example where the drawn circle represents at two levels simultaneously (as graph of the equation for a circle and as the image of an actual geometrical circle). Thus arises another disanalogy between the Minkowski spacetime interval and its exemplar the Pythagorean line element. Distances on a drawn right triangle are images of actual distances in a geometrical right triangle. Spatial distances on a Minkowski diagram, by contrast, are not likewise images of distances in spacetime, for a "world line" on a Minkowski diagram is a graph of the *equation* for the trajectory of a body in special relativistic space and time. That is why if I plot the trajectory of a light ray, for instance, and then measure the distance between two of its points on the graph in front of me, this distance does not represent anything at all. The spacetime distance for a light-like interval is zero, after all, while the spatial and temporal intervals traversed by the light are registered on the x and t-axes respectively.

The rather obvious distinction between a graph and an image would hardly be worth mentioning except for the propensity of physicists and philosophers of science to speak of spacetime diagrams as if they were images rather than graphs. Indeed, the difference between a graph and an image, for which adepts in relativity undoubtedly have an intuitive sense, is nevertheless systematically disregarded in some of the best literature on spacetime.[12] An example is Roberto Torretti's admirable *Relativity and Geometry*, a historical-critical analysis of special and general relativity "from the standpoint of geometry."[13] In his opening discussion of Newtonian physics, Torretti introduces the concept of a coordinate system or "mapping" of points in spacetime to real numbers. The variables x_n are introduced in purely analytic-algebraic terms without being associated with any set of coordinate axes or spacetime diagram. At a certain juncture, though, we are suddenly informed that a *curve* in the four-dimensional manifold M may be called a "worldline." Evidently we have made the transition to a graph without being explicitly told that we have. Surely nothing to complain about, except that further on in the same discussion we learn that "*a free particle has a straight worldline*: this is the geometric contents of Newton's First Law."[14] Strictly speaking, of course, it is not the geometrical contents of Newton's First Law, but rather the geometrical contents of a graph of the equation $\left(\frac{d^2x_i}{dt^2}=0\right)$ for Newton's First Law. Fortunately, just two lines later Torretti backtracks slightly: "The difference between a free or inertial and a non-inertial or forced motion is therefore *analogous* [my italics] to that between straightness and crookedness ..." (30). Perhaps the term "analogy" is less than apt here, though, suggesting as it does a degree of sameness in the order of being, whereas the relation between actual inertial motion and geometrical straightness on a graph is a purely symbolic or conventional. That is nothing against the use of spacetime diagrams, but if we are seriously to accept the proposition that an inertial trajectory is a "geometrical object," as writers on spacetime habitually insist, then such a trajectory should be literally geometrical—in this case literally a straight line. Or, if the notion of "straight line" is not to be taken literally, then we should be told what actual geometrical object is here intended.

Torretti's ambiguity is emblematic of a general propensity, with which the literature on spacetime is replete, to conflate images and graphs, direct representations with symbolic representations. An extreme example of this propensity is Petkov's contention that the law of inertia can be understood in terms of the four-dimensional "stress" experienced by a deformed (crooked) world tube, with inertia thereby understood as the force that restores the world-tube to straightness.[15] Needless to say, such a deformed world tube (crooked line) lives only on a graph and experiences no stress. Once again, we can always straighten out the deformed world-tube and relieve the stress by means of a suitable coordinate transformation.

Another example, again from one of the better studies, illustrates the same tendency to reify symbolic entities. In his estimable *Foundations of Spacetime Theories*, Michael Friedman introduces the notion of a vector tangent to a curve in Euclidean three-space R^3, with a "component-free" representation as in Fig. 3.2 below[16].

The curve may be regarded as an image or direct representation of the geometrical object of interest—a curve in space. The vector itself merits some additional comment, though, since it serves in one respect as an image and in another as a graphic symbol. Its "imaging" function consists in pointing in the same direction as the curve at the point of tangency. However, the length of the tangent vector is purely symbolic, since the rate of change represented by the length of the drawn line is a different kind of quantity from the length of the line drawn on the diagram. In this case we symbolically represent the magnitude of the rate of change by means of the length of a line. Thus we have an image (direction of the arrow) in one respect and a symbol (length of the vector sign) in another.

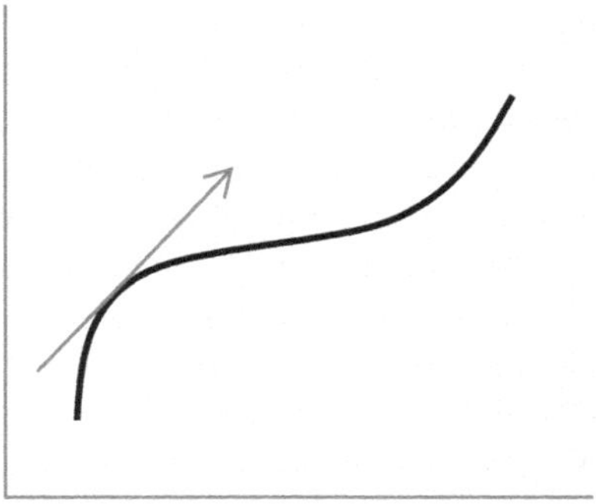

Fig. 3.2 Tangent vector in Euclidean 3-space (one dimension suppressed)

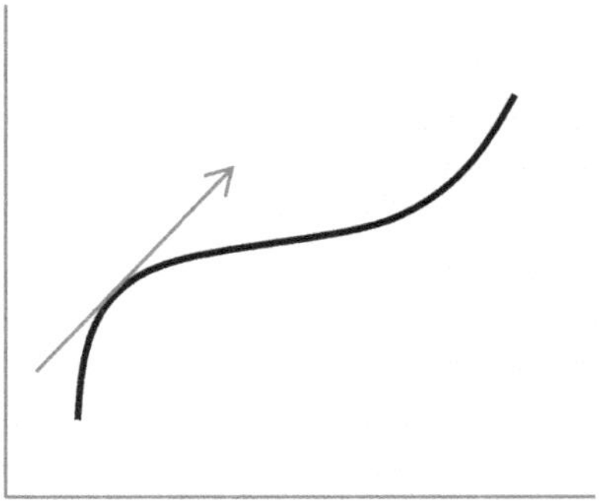

Fig. 3.3 Tangent vector in spacetime (two dimensions suppressed)

Friedman then seamlessly proceeds to an analogous representation of a curve and its tangent vector in the four-dimensional spacetime manifold R^4 as above.[17]

Figure 3.3 above is clearly the very same diagram as Fig. 3.2, except that the vertical axis now represents time. But notice what has happened. In the earlier example (Fig. 3.2), the geometrical object of interest, a *curve in space*, was represented by an *image*—a spatial curve on the diagram. But now our "spacetime curve" is purely symbolic. It contains no imaging features whatsoever. The physical trajectory in question is not actually a line, straight or otherwise, since the line is solely on the graph. Moreover, the direction of the arrow for the tangent vector no longer represents an actual direction in space, as before, but now symbolically represents the body's velocity. Furthermore, the length of the "four-vector" on paper is now completely meaningless. While Friedman emphasizes the superiority of the "component-free" mode of representation, which is supposedly more direct, dispensing as it does with coordinate systems, in reality the freedom from components is illusory and merely notational. Minkowski spacetime, after all, is a construction out of *Cartesian components*. The best we can manage as regards a component-free representation is to combine the three spatial components into a single "component-free" spatial component of our spacetime vector. But we simply have no means of combining the spatial component and the temporal component into a component-free entity or single vector, except merely notationally—by writing for the tangent vector in spacetime the symbol $\boldsymbol{T}$ instead of T^i, for instance—or graphically by drawing an arrow tangent to a curve on a spacetime diagram. That is why when so-called "coordinate-free" or "component-free" methods are used in four-dimensional physics, to actually do any calculation or measure anything one still has to restore at a minimum the distinct time and space components. In

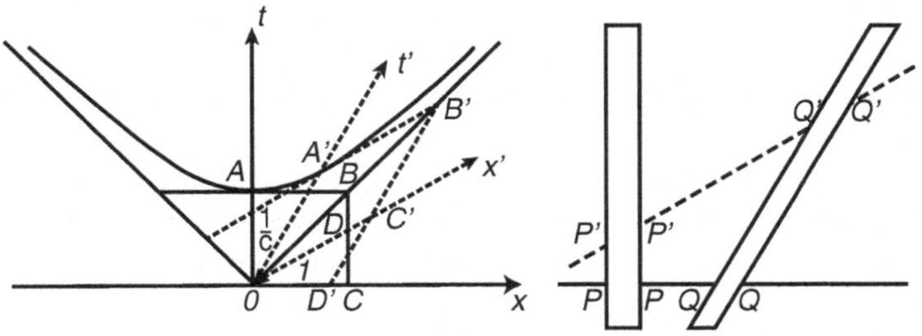

Fig. 3.4 Minkowski's spacetime cross-section

a genuinely component-free or vectorial formulation it would be entirely optional whether to resolve geometrical objects into components.

A last example from the spacetime literature helps bring home the invidious effect of mistaking graphs for images. Petkov argues, on the basis of Minkowski's cross-section diagram (see right hand part of Fig. 3.4 above), that relativistic length contraction would be impossible in a three-dimensional world.[18] For in that case the respective four-dimensional strips on the diagram allegedly would not represent anything existing in the world, with the result that there could be only the one proper length *PP* corresponding to the single projection onto the *x*-axis. But in Minkowski's absolute world, Petkov notes, the object can also have length $P'P'$ corresponding to its projection parallel to the x'-axis in the space of the relatively moving observer.

Petkov concludes, "Therefore the four-dimensional vertical strip of the body would not represent anything real in the world and would be merely an abstract geometrical construction" (34).

Clearly, the four-dimensional vertical strip *both* represents something real in the world and is merely an abstract geometrical object. That very duality, as we learned above, is the essential property of a graph. Petkov has unfortunately taken Minkowski's diagram for an image when it is in reality a graph of the points at which certain events occur in space and time. What Petkov more intelligibly could have said is that while the strips on the graph do not *image* anything in the physical world, and are indeed merely abstract geometrical constructions, they do indirectly or symbolically represent something in the real world, namely, the Lorentz contraction.[19]

Returning to Minkowski's hyperbolic rotation in spacetime, we must conclude that the analogy with a Pythagorean rotation in Euclidean space is merely formal. We cannot physically rotate a "time axis," after all, whereas if I wished I could set up a Cartesian grid in my backyard and verify the Pythagorean distance formula by actually rotating the apparatus. But the hyperbolic rotation can be effected only on a graph. It is a symbolic rotation and should not be taken as the image of a "Lorentz boost" in spacetime. There is no physical "angular rotation" whatever, although the construction does bear a limited algebraic analogy to the algebraic representation of a real rotation in space. Physicists and philosophers know this intuitively, of course, which raises the question of the source of the considerable seductiveness of the Minkowski diagram. That source, I shall argue in Part II, is the symbolic-algebraic constitution of the spacetime interval itself.

Ultimately, if our intention is to represent a four-dimensional geometrical object, we should be able to identify that object without recourse to symbolic representation on a graph. In pre-Minkowski special relativity we can easily translate the language of graphs, which is unobjectionable in itself, into something physically intelligible. That unfortunately does not appear to be the case for the theory of Minkowski spacetime.

3.3 Invariance and Special Relativity

According to the regnant account of the history of relativity, although Einstein discovered special relativity in 1905, the theory's true significance was opened up only in 1908 when Minkowski unveiled its "deep geometrical structure." Since Minkowski's theory countenances solely invariant geometrical objects, so goes the account, it alone discovers "objectively reality." If invariance truly merits the significance assigned to it in such an account it must speak strongly in favor of Minkowski's theory. If we consider the concept of invariance more closely, though, the proposition becomes dubious.

3.3.1 Invariance and Objectivity

The idea that invariance is an index of objective reality is one of the most deeply entrenched notions of relativistic physics. Scott Walter plausibly suggests that one of the principal reasons for the relatively rapid acceptance of Minkowski's theory, subsequent to its publication, was just the restoration of the kind of absolute entities that physicists were used to, but which had been relativized by Einstein's 1905 theory.[20] Yet Einstein himself, even after embracing Minkowski's formalism and employing it in general relativity, does not appear to have subscribed to this view of invariance.

The principle of relativity as Einstein understood it, after all, holds that the *laws of nature* are invariant, not that particular physical quantities or geometrical objects are invariant. Indeed, in Einstein's thought the concept of *covariance* brings out this very difference: quantities such as length, mass, and so forth must *co-vary* so that the laws of nature may remain invariant. However, most if not all of the influential interpretations of relativity over the past few decades have embraced invariance, or better *frame-independence*, in preference to *relativity*, as the ruling concept of spacetime theory. The doctrine evidently has its historical origins in Felix Klein's "Erlanger program," where differing geometries are classified based on their invariants under some group of transformations. It finds its way into mathematical physics via Minkowski's geometrical interpretation of the special theory of relativity.

Considered in itself, of course, special relativity is an abstraction, for the special theory holds only infinitesimally in general relativity. We must think of special relativity in local terms and, specifically, refrain from extrapolating results that depend on frame-relative simultaneity at a distance. Nevertheless, the notion of frame-relativity itself remains essential to Einstein's 1905 special relativity. Nerlich, one of the most intelligent defenders of invariance theory, argues that apart from Minkowski's four-dimensional geometry, the three-dimensional "rod-in-a-frame" of Einstein's 1905 special relativity is referentially indeterminate: "In the 1905 theory why don't know quite what we are talking about."[21] The problem is that the pair of simultaneous events determining the length of the rod is not the same in two frames moving relative to one another. Thus, at least if a rod is defined as a set of simultaneous events, we have two different rods rather than one and the same rod measured in two different frames. A referentially determinate rod, then, evidently would be a four-dimensional entity whose frame-independent proper length measures the space-like interval between events at its endpoints. However, it is only in the four-dimensional theory itself that we fix the identity of a three-dimensional rod in terms of a set of simultaneous events or a three-dimensional cross-section of the rod's four-dimensional "world tube." In Einstein 1905, the three-dimensional rod is taken for granted as an enduring object with a proper length in its rest frame, apart from considerations of simultaneity. This assumption is perfectly reasonable, for we cannot even begin our reflections in the theory of relativity without assuming the existence of entities that maintain their identity through time. A mere manifold of events could never yield the world tubes of the four-dimensional theory itself. With respect to the rod, a single pair of simultaneous events is required, not to determine the rod referentially, but to measure its length when it is moving relative to a rest frame. Thus is the one and only three-dimensional rod contracted in a frame, with no issues of referential indeterminacy such as concern Nerlich. Length contraction is a physically real effect in special relativity, apart from considerations of relative simultaneity at a distance.[22]

What makes the invariance theory seductive is a misplaced analogy with geometry. Suppose I am describing a geometrical object by means of a rigid reference frame, for instance, a Cartesian coordinate system. If I reorient the reference system, its relations with the object of interest change. These changed relations are entirely objective—they do not reflect an observer's "point of view" or anything like that. However, given that my aim is to describe the object of interest itself, not anything else, I do not want the changing relations to the reference system to enter the description. In this sense I am indeed seeking invariants in coordinate geometry: the distance between two points of interest, for example, irrespective of the changing components of displacement vector between those points. Observe that there is nothing in the *invariant*, in this case the distance between the points of interest, which makes it "objectively real" in some way the changing vector components are not objectively real. To be sure, the invariant distance between the points is "objective" in the sense of being the object of interest. But both the distance between the points and the vector components are objectively real, even if the latter are not invariant because I am deliberately altering them. Since I am the one using this particular physical system as a reference frame, though, I might well say that "relative to me," or from my perspective, the invariant displacement vector between the two points has different components than it does from the perspective of somebody else employing a different reference system. But this is only a loose way of speaking, since the vector components themselves are not geometrically related to me as an "observer," but rather to the reference system itself, a physical system made of rigid rods. I just happen to be using it, and nothing would change if I were not.

The success of the "method of invariance" above depends on the rigid rods staying the same length when the coordinate system is reoriented; for if the rods changed their length, then the displacement vector itself, and not just its components relative to the reference system, would be altered. In that case, we would have to either disqualify the reoriented coordinate system or incorporate it into our description of the geometrical object of interest. On the latter option, length would now be a relational property of the displacement vector with respect to the reference system. That is, the determination of length would now require specification of a frame, just as, for instance, in the description of left and right when I say the coffee cup on my desk is "to my left," while at the same time it is to your right. There is no question here of any loss of objectivity that previously attached to the invariant distance between points. Relational properties are as objective as non-relational ones. The displacement vector in the new situation would have a definite and objectively determined length, but only relative to a frame.

The significance of invariance in geometry lies not in serving as an index of objective reality, but rather in picking out the properties of the geometrical object under study and screening out other properties, equally objective, which depend merely on the object's relations to the reference system. Philosophically speaking, the opposite of "objective" is *subjective*, not *relative*. But in special relativity, unlike geometry, we have to take into account the effect of motion on the rods and clocks of the reference system. Therefore, we can no longer use invariance the way we did in geometry. The length of a rigid rod is now a relational property of the rod, and similarly for clocks, which now register a time interval relative to a frame.

Reflecting in 1920 on the problem of the magnet and conducting coil, famously mentioned at the beginning of his 1905 paper, Einstein observes,

> The idea that these two cases [motion of the magnet or motion of the coil] should essentially be different was unbearable to me. According to my point of view, the difference between the two could only lie in the choice of the point of view, but not in a real difference .... As seen from the magnet, there was certainly *no* electric field; whereas seen from the circuit there certainly was an electric field. Therefore, the existence of the electric field was a relative one, depending on the state of motion of the coordinate system used; and only the magnetic and electric fields *combined*, aside from the state of motion of the observer or coordinate system, could be granted a kind of objective reality.[23]

In light of our analysis, Einstein expresses himself unfortunately when he suggests that only the combined field is "objectively real," as if the frame-relative electric field merely reflected a subjective point of view. Should we characterize the objective reality of the combined field this way, in terms of its independence of the motion of the "observer or coordinate system," as if the latter two were equivalent? As we said, the coordinate frame is a physical system comprised of rigid rods and clocks; an objective state of affairs holds relative to that system (how fast the magnet is moving, for instance) quite apart from the presence of any observer. Conversely, observers can use any coordinate system they like and are not "assigned" to their rest frames. To be sure, it is harmless to speak of reference frames in terms of observers at rest in those frames, as long as we do not also attribute to these physical systems themselves the subjective perspective of observers. The frame-relativity of the electric field has nothing to do with how things subjectively appear to observers at rest in this or that frame, and Einstein in fact leaves it unclear the nature of the combined field in the magnet and coil scenario. Does the combined field act through physically distinct electric and magnetic components? How so if those components are objectively unreal? And if the components are not physically real, then out of what elements is the field "combined"?

It is just at this point that spacetime geometry is supposed to come into play, furnishing an objective description of the combined field as a single four-dimensional object. Before treading the geometrical path, though, we should reflect more carefully on the question of frame-relative forces in special relativistic electrodynamics. A force is either exerted or not, regardless of reference frame, but to quantify a force we need to consider its effect. A force in the special relativistic version of Newton's second law, $f = m_0 a\gamma$, has for its effect frame-relative increments of both kinetic energy and momentum in the accelerated body. We learned in Chap. 2 that the Minkowski four-momentum does not in fact combine relativistic kinetic energy and relativistic momentum into a single frame-independent quantity, since the time component of the four-momentum is $mc\gamma$, not $mc^2\gamma$. The four-dimensional approach fails here, and if it fails in this case it fails in every case. Kinetic energy is the ability to do work, and work is a motive force acting through a distance. By virtue of the Lorentz contraction, a body in motion is not capable of the same amount of work in every frame, since the distance through which the motive force acts is not the same in every frame. The motive force itself is either present or not, either acts or does not act, irrespective of the frame of reference; but the components of the physical effect fall out differently in different frames of reference.

For the case of the magnet and conductor, of course, the motive force itself acts through electric and magnetic components. However, a better description than the one Einstein gives in 1920, perhaps, would be that the components of the "combined field," which is either present or not, are relational manifestations of the one electromagnetic force. They are not objectively unreal for being relational. Clearly, the electromagnetic field is a dynamical effect of the relative motion of the conductor and magnet, not a dynamical effect of the motion of either of those relative to the reference system. It is solely the metrical properties of space and time that are determined relative to the reference system.

This is the solution to the case, cited by Nerlich, of a magnet and coil at rest relative to one another. Relative to a frame in which the magnet and coil are at rest, there is no induced electric field and therefore no current in the coil, while relative to a frame in which they move, according to Faraday's law, there is a magnetically induced electric field and therefore a measurable current. However, the motion of the electrical charges in the magnetic field produces an electromotive force which exactly cancels out the induced electric field, such that no current flows in the coil. Surely, as Nerlich puts it, "No coil is both devoid of current-producing **E** fields and yet contains two such equal and opposing **E** vector fields."[24] That is indeed the case, for in the scenario under consideration there are present no **E** fields, since there is no relative motion between the magnet and the coil. Nevertheless, the Maxwell-Hertz equations cited by Einstein in the electrodynamical part of his 1905 paper predict an induced electric field relative to the "stationary frame."[25] That is because these equations determine the existence of the electric field based on the motion of the charge relative to a rest frame, not relative to the coil. Accordingly, the opposing electromotive force relative to the reference system must also be considered, as if both forces were present.

But Einstein's essential insight in 1905 was that it is the motion of the magnet and coil relative to one another that gives rise to the dynamical effect of the field, not the motion of either of those relative to a rest frame. (He misstates that insight when he says that only the combined field is "objectively real.") In reality, the motion of the charge relative to the reference system determines only the metrical environment in which the force acts, so to speak, not the dynamical production of the force itself. As Einstein writes in 1905,

> If a unit electric point charge is in motion in an electromagnetic field, the force acting on it is equal to the electric force which is present at the locality of the charge, and which we ascertain by transformation of the field to a system of coordinates at rest relatively to the electrical charge. (New manner of expression.)[26]

This procedure gives the **E** field due to the relative motion of the charge and the magnet, in the present example zero. The problem is that if we transform to any frame in which the charge is moving, the Maxwell-Hertz equations will return an electromagnetic field that is not really there, since in pre-relativistic fashion the Maxwell-Hertz equations assume that the field is generated by motion relative to a rest frame. The fact that in special relativity the rest frame is no longer the ether, but rather any inertial frame, leaves the anomaly intact: for it is not motion relative to a rest frame that gives rise to the field, but rather the relative motion of the magnet and coil. Four-dimensional spacetime does not remove the anomaly. Instead, once we have determined the field in the rest frame of the charge, we can apply $f = m_0 a\gamma$ to obtain $\mathbf{E}' = \mathbf{E}\gamma$ for a frame relative to which the charge is moving. For Nerlich's case of the magnet and conductor at rest relative to one another, the electric field is zero in any frame. For other cases we obtain from $\mathbf{E}' = \mathbf{E}\gamma$ a physically real, frame-relative electric field.

Nerlich reasonably stresses the difference between *frame-invariance* (so the velocity of light, for instance) and *frame-independence* (for example, how many passengers are riding on one of Einstein's famous trains). What supposedly renders true invariants objectively real is not that they are the same relative to all frames, but that they are independent of any frame at all. It is only because such objects are themselves absolute and frame-independent that they can appear in our experience as invariant. On such a view, then, *frame-invariance* must be regarded as the mode in which that which is absolute reveals itself to frame-bound beings like us. This philosophy is often supplemented by the assertion that the absolutes of a theory are therefore its "observables"—for instance, in relativity the "proper" quantities such as proper time, proper mass, and so forth. Thus the proper time registered by a clock, for instance, is to be understood not as a measure of the time elapsed relative to the frame in which the clock is at rest, but rather as a measure of the absolute spacetime distance traversed by the clock.

On its face, the observation that proper time is an "invariant" in the theory of relativity seems trivial, as if one were to read great significance into the fact that all observers can agree that the clock face has Arabic rather than Roman numerals or something. For the very designation of the time recorded on a particular clock as the "proper time" renders that time frame-independent by definition—it is, after all, the time recorded on *that* clock. This is not the same as to say that the quantity of time that has passed is frame-independent. Suppose I am driving down the highway and mark the reading on my speedometer as I pass a tree on the side of the road. The speedometer reading itself, the position of the indicator, is frame-independent and thus "the same for all observers." Does that mean that the speed of my car is frame-independent? Clearly not, since the speedometer registers my car's speed relative to the pavement.

Proper quantities might reasonably be regarded as frame-invariant in the sense that they serve as reference quantities in all frames. That is, the coordinate time in every frame is $\tau\gamma$, and so the proper time τ itself is frame-invariant. But τ is still the time registered in the clock's rest frame and, in any event, this sense of frame-invariance is quite apart from absolute geometrical objects. To be sure, in special relativity we cannot regard the proper time recorded on an *accelerated* clock as the time of the clock's rest frame. But that is simply because the rest frame of an accelerated clock is not an allowable frame in special relativity. Such a clock does not register an amount of time that has passed in any frame.

What really motivates the claim that proper time is a frame-independent absolute in the theory of relativity is the idea that a clock records the absolute spacetime distance along its path. That is not true. To be sure, proper time is *proportional* to the Lorentz-invariant $\sqrt{c^2t^2 - x^2}$, the so-called "interval" in spacetime (as per to the relation $c^2\tau^2 = c^2t^2 - x^2$). Unfortunately, this proportion is often erroneously interpreted to mean that the proper time registered on a clock actually "measures" a spacetime interval. Such a manner of speaking is misleading, however, since a quantity can properly be *measured* solely in terms of a unit homogeneous with the quantity itself. If I wish to measure how much money I have, for instance, the unit of measure must be an amount of money. Of course, I may *represent* the amount of money I have by means of some other quantity, the length of a line, for instance, but that does not amount to measuring my wealth in units of length. Likewise, the fact that the square of the proper time registered on a clock is proportional to the invariant algebraic quantity $c^2t^2 - x^2$ does not mean that $c^2t^2 - x^2$ is a frame-independent physical quantity measured by the clock. After all, mvc^3 is proportional to momentum, even though mvc^3 obviously represents no physical quantity at all in the real world. Moreover, while the quantity $c^2t^2 - x^2$ is clearly frame-invariant, it most assuredly is not *frame-independent*. Quite the contrary, the expression is meaningful only as one member of a Lorentz transformation equation. A truly frame-independent expression for the spacetime interval would have stand alone, like the Pythagorean distance formula, independently of any transformation law. In no wise, then, can proper time be regarded as "frame-independent" simply by virtue of its proportion to the Minkowski interval $\sqrt{c^2t^2 - x^2}$.

If we are committed Minkowskian absolutists, of course, we will insist that proper time cannot be a relativistic quantity because relativistic quantities are not objectively real; but if we are relativists we will demur and continue to regard proper time in special relativity as relative to the rest frame of our clock, and proper length as relative to the rest frame of our rigid rod. And whether we should be absolutists or relativists is a philosophical question that cannot be settled by relativistic physics. The coffee cup on my desk is to my left and to your right. It is precisely as participant in these relations that the coffee cup is "referentially determinate," in Nerlich's terms. The relations in question are objective and have nothing to do perspectival cross-sections or anything of the kind. As for special relativity as an empirical theory, it is enough that a length-contracted rod exhibit sufficiently stable relations with other entities to secure its referential determinacy. For if the length-contracted "rod-in-a-frame" were not the same rod as the non-contracted "rod-in-its-rest-frame," on what basis could we say that this particular and perceptually given rod is "really" a four-dimensional entity in Minkowski spacetime?

3.3.2 Invariance and the Clock Paradox

There exists a growing body of literature suggesting that the so-called relativistic "clock paradox" (or "twin paradox") can be explained solely in terms of the invariance properties of Minkowski spacetime.[27] To review the scenario, two clocks are locally synchronized initially, with one of them subsequently traversing a closed path in space and finally reuniting with the "stationary" clock at some later time. At the time of their reunion the traveling clock is retarded compared to the rest clock. The evident paradox lies in the failure of relativistic reciprocity: for the traveling clock is retarded absolutely compared to the rest clock. This failure of reciprocity has suggested to some authors that clock retardation is not a relativistic effect at all, but rather an absolute effect determined by the four-dimensional path of the traveling clock, which traverses a shorter interval in spacetime than the stationary clock.

Petkov, for instance, maintains that philosophical considerations aside only the Minkowski four-dimensional viewpoint is consistent with the physics of the clock paradox.[28] To bracket considerations of acceleration that are irrelevant to the clock paradox, Petkov employs an all-inertial clock scenario with a third clock serving as surrogate for the original traveling clock on its return trip. That is, traveling clock B departing from A is intercepted by synchronized clock C traveling at equal and opposite velocity, such that when C meets A, C registers the same retardation compared to clock A as accelerated clock B would have in the original scenario. Thus no acceleration is involved.

The problem with the three-clock scenario, though, is that it is not set up to register reciprocal clock retardation. To exhibit the expected reciprocity we must reanalyze the scenario in the rest frame of clock B. The extra clock therefore must function this time as a surrogate for A, just as in the original scenario it stood in for B. However, C fails to reunite with B after to C's encounter with A; furthermore, in B's frame C is now traveling at twice A's speed and so the required symmetry between A and C would not be satisfied in any event. Thus, in the three-clock scenario analyzed relative to the rest frame of B, there is available no direct comparison for determining the retardation of clock A compared to clock B. If we suitably modify the three-clock scenario with a new surrogate inertial clock D, standing in for A at the reunion with B, we shall find the expected reciprocity in time dilation.

As for the acceleration of the traveling clock in the standard twin scenario, Petkov is correct that it is not acceleration per se the yields the retardation of the traveling clock. Nevertheless, the accelerated clock will always be retarded compared with the non-accelerated clock. That follows simply from the fact that the validity of special relativity is restricted to inertial frames. In the usual scenario, after all, the clock that moves must be retarded relative to the rest clock; and in order for there to be a reunion at all, the moving clock must turn around and come back (that is, it must accelerate). Thus the accelerated clock will always be the time-dilated clock, an absolute effect to be sure, but no paradox because reciprocity in time dilation assumes a situation analyzable in terms of inertial frames in motion relative to one another. The twin paradox is therefore resolved by a standard analysis in special relativity, formulated three-dimensionally, with no necessary appeal to the "length of a world-line" in four-dimensional spacetime.

3.4 On the Arrow of Explanation

In his 1905 paper, Einstein removes the apparent conflict between the special principle of relativity and the light postulate by means of his analysis of simultaneity and the subsequent derivation of clock retardation and the Lorentz contraction. The slowing down of clocks and the shortening of rigid rods exactly compensates for the relative motion of reference frames, leaving the speed of light unaffected. However, Einstein does not claim to have explained why these effects occur, and any attempt to regard his presentation as an explanation must end up in circularity. For should we say that clocks slow down and rods contract because the speed of light is constant in all frames, we obviously are in need of an explanation for why the speed of light is constant in all frames, which could only be that clocks slow and rods contract. Supporters of Minkowski's theory customarily invoke the geometry of four-dimensional spacetime as a "constructive explanation" for relativistic time dilation and length contraction.[29] But even if we accepted Minkowski's theory, it could supply no more explanation of these effects than does Einstein 1905, for the four-dimensional geometry in question is derived from the Lorentz transformation in the first place. To provide a truly constructive explanation, the theory of Minkowski spacetime would have to account for the light postulate itself—why the speed of light is the same in all inertial frames. But Minkowski's theory presupposes rather than explains the light postulate.

Would length contraction and time dilation be better understood as dynamical effects, as John Bell and more recently Harvey Brown argue?[30] Einstein's 1905 special relativity, Brown emphasizes, was not generally interpreted at first as foreclosing the possibility of a dynamical account of relativistic effects such as length contraction and time dilation. But the acceptance of Minkowski's theory changed things:

> Following Einstein's brilliant 1905 work on the electrodynamics of moving bodies, and its geometrization by Minkowski which proved to be so important for the development of Einstein's general theory of relativity, it became standard to view the Fitzgerald-Lorentz hypothesis as the right idea based on the wrong [dynamical] reasoning.[31]

Such a dynamical account, Bell had suggested, might explain length contraction in terms of the equilibrium of electromagnetic forces between the particles of a rod in motion. However, it is hard to see how an explanation in terms of force could account for the reciprocity of the relativistic effects in question. Of two rigid rods, each shortened in the other's rest frame, only one of them, on Bell's hypothesis, has experienced an equilibrium-altering force. For the proposal to succeed we would require a sense of "dynamical" distinct from the concept of force.

One confusion in the discussion, as Brown notes, is the assumption that a phenomenon must be either "kinematical" or "dynamical."[32] The issue is not merely one of semantics. Brown points out that while Einstein placed his discussion of length contraction and time dilation under the "Kinematical Part" of his 1905 paper, Einstein was in fact cognizant of the fact that the concept of the "rigid" measuring rod and clock was not independent of dynamical considerations:

> Einstein realized, possibly from the beginning, that the first 'kinematic' section of his 1905 paper was problematic, that it effectively rested on a false dichotomy. What is kinematics? In the present context it is the universal behavior of rods and clocks in motion, as determined by inertial coordinate transformations. And what are rods and clocks, if not, in Einstein's later words, 'moving atomic configurations'? They are macroscopic objects made of micro-constituents—atoms and molecules—held together largely by electromagnetic forces.[33]

The kinematical-dynamical dichotomy is misleading for another reason as well, which is that what makes a kinematical description *kinematical* is that it abstracts from the causes of motion, including the absence of forces, not that it describes a "kinematical phenomenon." The medieval mean speed theorem, for instance, which abstracts from the cause of uniformly accelerated motion, does not thereby explain uniformly accelerated motion as a kinematical phenomenon. Even the law of inertia, which deals with unforced motion, is not merely a kinematical law. That is, we do not abstract from the forces causing inertial motion, since there are none. The law of inertia does not abstract from the concept of force itself, as does the mean-speed theorem, since the very statement of the law of inertia refers specifically to the absence of forces. The law of inertia is dynamical, and indeed no physical law is merely kinematical. That is why we characterize Einstein's metrical theory of gravity, with its appeal to the "most inertial" trajectory of a force-free body, as a *dynamical* theory of gravity rather than a kinematical theory of gravity. There are a number ways we might think of unforced motion as dynamical or caused—a body's endeavor to persist in its state, for instance, which is how Leibniz regarded it. A dynamical or causal explanation need not be framed in terms of forces.

We do not yet possess an adequate explanation of length contraction and time dilation, and I have none to offer here. Such an explanation would have to account for why these effects occur in the precise degree necessary to render the speed of light invariable in all inertial frames. It would be a vanishingly unlikely coincidence, comparable to the equality of inertial and gravitational mass in Newton's theory, for the relativistic factor γ to exactly compensate for the relative velocity of frames and so maintain constant light velocity in all frames. The situation indeed suggests that light, or at least light velocity, plays a role in the production of the rod and clock effects, perhaps functioning as a universal metrical standard in nature. I shall refrain from speculating on what the mechanism for this might be.

3.5 Conceptual Difficulties of Minkowski Spacetime: The Need for a Historical Approach

Based on our analysis thus far in Part I, the theory of Minkowski spacetime falls considerably short in conceptual intelligibility and coherence, however superior it may be in terms of mathematical formalism. Defenders of Minkowski's theory assure us that its absolute geometrical objects are physically real, but the theory itself fails to identify any such objects. Minkowski's formalism, with its "automatic covariance," undeniably represents a mathematical contribution to special relativity; but that is something else than to say it makes a real contribution to the theoretical content of special relativity.

While the evidence against the physical reality of Minkowski spacetime is thus far very compelling in my judgment, the argument is not yet conclusive. Virtually any scientific concept is infected with at least some degree of incoherence, and when a concept plays a vital role in our best theories, as would appear to be so for Minkowski spacetime at least in *general* relativity, if not special relativity, then rather than rejecting it outright we allow the process of scientific advance gradually to purge its incoherencies, at least as far as possible. Moreover, from the perspective of the kind of concept formation that characterizes modern science in general, we have perhaps focused too exclusively, for the case of Minkowski spacetime, on the particular obscurity of the concept's relation to what it intends. That is to say, I have found fault principally with the intentional structure of this concept or its manner of referring to the physical world it is supposed to be "about." But concept formation in modern science has from its inception in the sixteenth and seventeenth centuries been determined principally by the relation of concepts to *other concepts* in the system of science as a whole, not by the relation of concepts to what they intend per se. Indeed, this very feature of modern concept formation emerges very early on in modern mathematics, in the development of the symbolic concept of number from Vieta to Descartes. Jacob Klein characterizes the shift from the ancient to the modern mode of concept formation as follows:

> In Greek science, concepts are formed in continual dependence on "natural," prescientific experience, from which the scientific concept is "abstracted." The meaning of this "abstraction," through which the conceptual character of any concept is determined, is *the* pressing ontological problem of antiquity; it becomes schematized in the medieval problem of universals, and, in time, fades away completely. The "new" science, on the other hand, generally obtains its concepts through a process of polemic against the traditional school science. Such concepts no longer have that natural range of meaning available in ordinary discourse, by appeal to which a truer sense can always be distinguished from a series of less precise meanings. No longer is the thing intended by the concept an object of *immediate* insight. Nothing but the internal connection of all the concepts, their mutual relatedness, their subordination to the total edifice of science, determines for each of them a *univocal* sense and makes available to the understanding their only relevant, specifically scientific, content.[34]

If Klein is right that the concepts of modern natural science—Minkowski spacetime or any other concept—achieve their relation to what they intend *indirectly*, via the "total edifice of science," then we cannot fairly subject the concept of spacetime to a standard of intelligibility derived from the quite different intentional structure of the concepts of Greek science, where concepts intend their objects *directly*. Cleary the concept of Minkowski spacetime is accepted today principally by virtue of its role in the formulation and conceptual structure of Einstein's theory of gravity, not by virtue of the transparency of its relation to what it intends.

If we wish to more than merely acknowledge the incommensurability of the ancient and modern modes of concept formation highlighted by Klein, however, we need a method of evaluating the concepts of modern mathematical science on their own terms, but without simply taking for granted the superiority of the modern mode of concept formation. This can be achieved if we recognize that the original concepts determining modern science are in their very meaning-structure transformations of received pre-modern concepts, and specifically of Greek concepts. Thus, to grasp the present-day meaning of a concept in the total edifice of science, one must view that total edifice—within which alone meaning accrues to the concept—from the perspective of its historical constitution.

A historical approach such as described above is all the more indispensable for elucidating the symbolic-algebraic structure of the concepts of modern mathematical physics. For there can be no doubt that this very structure first arose as a transformation of the received Greek concepts of ratio and proportion, traditionally regarded up to the time of Newton and beyond as the proper mode of expression for a mathematical science of nature. Our analysis suggests as much already in some of the confusion we have noted in the spacetime literature regarding the physical meaning of graphed algebraic equations. This very mode of representation, which makes its first appearance in Descartes' *Geometry* of 1637, has somehow determined the conception of mathematical physics ever since.

Notes

1. This observation applies, for instance, to the so-called "color manifold": If we characterizing each possible color in "color space" in terms of three variables (for instance, hue, brightness, and saturation), we can define a "color metric" measuring how different one color is from another. However, we do not thereby attain a single continuum, since the color metric governs the color manifold solely as a *symbolic space* (or configuration space). The three "dimensions" (hue, brightness, saturation) remain heterogeneous. As Kinsman et al. observe "The only way for D [distance] to be a valid metric distance [in the color manifold] would be if [all] axes had the same units" (Kinsman et al. 2012).
2. Bohm 1996, 148.
3. Meyerson 1985 [1925], 72.
4. For one thing, since time has only one dimension we cannot highjack an extra dimension for the representation of space. Beyond that, space is present all at once, whereas time is present solely moment by moment and is for that reason ill-suited for representing anything else.
5. Minkowski 1952 [1909], 83.
6. Elie During (During 2012) emphasizes that the depiction of multiple reference frames on a single drawing is the decisive feature of the Minkowski diagram, which distinguishes it from a mere graph in the usual sense. However, During unnecessarily complicates his analysis by equating multiple reference frames with multiple "perspectives." While it is true that

one does occupy a particular perspective from the reference frame in which one is at rest, the concept of a reference frame in the theory of relativity is entirely distinct from that of an observer's "perspective," and any observer can use any reference frame.

7. Emily Grosholz (Grosholz 2007) distinguishes three overlapping forms of mathematical representation: iconic, symbolic, and indexical. Icons resemble what they represent, symbols represent by convention, and indexes "represent for the sake of organization and ordered display" (25). What we call a graph would be an index in Grosholz' terminology. John Roche (Roche 1993, 197–198) distinguishes between diagrams and illustrations, the former making no attempt at resemblance as in the representation of a body's length by the length of a line. Thus diagrams become conventional or symbolic when what they represent is no longer spatial, as when we represent a body's weight by the length of a line.
8. Roche (1993, 216) notes that a graph in the strict sense of the term would exhibit a set of experimental data, whereas what we often find instead is the graphical representation of an idealized law or equation.
9. The production costs for such illustrations were evidently very high in Newton's time, so he used the same illustration in both of the aforementioned lemmas.
10. Rule 14 of *Rules for the Direction of the Mind* (Descartes 1985–1991 [ca. 1628]: 1:65).
11. Grosholz (2007, 167) specifically highlights Descartes' definition of multiplication, according to which the product of a line by a line yields another line (rather than a plane). This innovation, which renders geometrical magnitudes symbolic, removes a traditional impediment the geometrical representation of arithmetical operations, since Descartes is longer limited to three dimensions.
12. According to Roche, this propensity has its origin with Minkowski himself: "Lagrange, Laplace, Poinsot and their followers would surely have regarded Minkowski's space-time as a conventional configuration space [symbolic space] created for purposes of mathematical convenience. ... Minkowski, however, had far more ambitious claims for his invention. ... From the perspective of the present article Minkowski seems to have taken his geometrical metaphor of space and time rather too literally" (Roche 1993, 231).
13. Torretti 1983, 3.
14. Torretti 1983, 29.
15. Petkov 2005, section 10.2.
16. Friedman 1983, 35. I have redrawn Friedman's diagrams for purposes of this discussion.
17. Friedman 1983, 36.

18. Petkov 2012, 33–34.
19. I have here omitted from consideration Petkov's further argument that, save for the Minkowski's four-dimensional viewpoint, we are actually dealing with two different three-dimensional objects in their respective frames. That may indeed present a problem for a so-called presentist interpretation of time. However, in the present discussion I am proceeding with no such presentist assumptions, but rather evaluating the concept of Minkowski spacetime on its own terms. To be sure, "presentism" is problematic from the perspective of special relativity, with or without Minkowski spacetime. But special relativity is valid only infinitesimally in general relativity.
20. Walter 1999, 72.
21. Nerlich 2013, 89.
22. See Winnie 1970, 97–98.
23. Einstein 2002 [1920], 135.
24. Nerlich 2013, 88.
25. Einstein 1952a [1905], 51–52.
26. Einstein 1952a [1905], 54.
27. See for example, Arthur 2006, sections 3–4; Nerlich 2013, Chap. 7; Petkov 2005, section 5.5 and 2013, 86–88.
28. Petkov 2005, section 5.5.
29. See, for instance, Janssen 2009, Balashov and Janssen 2003, Nerlich 2013 (Chap. 5). On the other side of the debate is, most prominently, Harvey Brown (Brown 2005; also Brown and Pooley 2006).
30. Brown 2005; Bell 1987.
31. Brown 2005, 2.
32. Brown 2005, 4.
33. Brown 2005, 4.
34. Klein 1992 [1934–1936], 120–121.

PART II

The Symbolic-Algebraic Constitution of the Concept of Spacetime

1.1 Introduction to Part II: The Concept of a Sense-History

We have noted that Cartesian algebra was first introduced into mathematical physics in the seventeenth and eighteenth centuries as a symbolic translation of Euclidean proportions. Of course, the historical observation does not in itself entail the further claim that Euclidean proportion determines the present-day meaning of algebraic equations in theories of physics. Indeed, we would fall prey to what is sometimes called the "genetic fallacy" if we assumed that the historical origin of a concept rules its subsequent meaning development. Instead, we must distinguish between a contingent chronology of historical influence, so to speak, and what we should call a historically inscribed logical structure.

Edmund Husserl's later philosophy of science focuses on just this question of the genetic constitution of meaning and the associated role of historical investigation in the clarification of concepts. Moreover, Husserl's analysis of what he calls "Galilean science" has specifically in view the very formalism-inspired Göttingen physics of which Minkowski is so prominent an exemplar.[1] Husserl and Minkowski were colleagues at Göttingen from 1902 until Minkowski's death in 1909 and Husserl was well acquainted with Göttingen science. I therefore base my account of the concept of "sense-history" principally on Husserl, especially his "The Origin of Geometry" of 1936.[2]

The sense-history of a concept is logical, not temporal or "historical" in the usual sense, even though historical investigation may be essential to our knowledge of a sense history. The concept of multiplication, for instance, presupposes addition, since to multiply a quantity is to add it repeatedly to itself. We can construct an algebraic concept of multiplication upon this original sense of repeated addition, such as multiplying a negative number by another negative number according to the rules of a symbolic calculus, but such a development of the meaning of multiplication is necessarily founded—not just temporally but also logically—upon the original sense of multiplication as repeated addition. That is, the algebraic concept of multiplication carries within itself, as a constituent of its present-day meaning, its original sense as repeated addition. In general terms, according to Husserl,

> [t]he essential peculiarity of such products is precisely that they are senses that bear within them, as a sense implicate of their genesis, a sort of historicalness; that in them, level by level, sense points back to original sense ...; that therefore each sense-formation can be asked about its *essentially necessary sense-history.*[3]

Thus, while the sense-history of a concept or a judgment is in itself logical rather than temporal in structure, it nevertheless presupposes a temporal genesis in the concrete process of actual history. The modern algebraic concept of multiplication, which carries within itself as part of its sense-history the ancient Greek concept of multiplication, did not arise until around the sixteenth century.

A related consideration is the inevitable historical process through which historically layered moments of sense become successively hidden from view. That is to say, in the historical sense we "forget" that repeated addition is still active in and continues to determine the present-day meaning of multiplication. In this sense, we should say that the concept of repeated addition is historically "sedimented" in the concept of algebraic multiplication. A primary goal of Husserl's form of analysis, then, is to "desediment" such meaning formations, as it were, and so reactivate the original experiential or intuitive evidence upon which they are ultimately grounded.[4] The outcome of the enterprise, if it successfully opens up the sedimented history of a concept, will be one of the following: (1) an adequate intuition of the object itself, (2) a dissolution of meaning, or most likely (3) a residue of coherent content.

Husserl observes that a "handed-down" science like the European science of his time constitutes itself through a tradition, and so the sense-structure of that science must be revealed through a historical investigation of that tradition.[5] The point is evident enough if we merely consider the fact that the term "science" as we use it nowadays refers specifically to "modern science" or the science handed down since the scientific revolution of the sixteenth and seventeenth centuries. In the present context this "ready-made" science must be further specified as *algebraic physics.* In any such tradition, each development has reference to a prior acquisition, and each new acquisition is taken up into the totality of the tradition as a premise for further acquisitions. In general, any scientific tradition as "ready-made" is to a significant extent passively received, in the sense that we do not begin from new foundations with each generation, but accept the accomplishments of the past more or less without question. This is a necessary feature of any scientific tradition and applies to even the most creative scientists. Furthermore, and this is the crucial point,

> since meaning is grounded upon meaning, the earlier meaning gives something of its validity to the later one, indeed becomes part of it to a certain extent. Thus no building block within the mental structure is self-sufficient; and none, then, can be immediately reactivated by itself.[6]

In terms of Minkowski's theory, we cannot gain possession of the concept of "spacetime four-vector," for example, without first having acquired "spacetime displacement," the latter itself founded upon the prior acquisition of the Lorentz transformation, and so forth. A passively acquired meaning *at any point* in the historical genesis of a concept thus may render the concept itself passive in the present-day as such accretions of meaning are successively sedimented in the concept.

The passive kind of thinking described above thus yields what we might call ready-made concepts, constituted by means of a sedimented sense-history. What we learn in scientific textbooks, essentially, is how to operate with such ready-made concepts in a rigorous and methodical way. To render transparent the present-day meaning of such concepts thus requires a form of historical investigation we could call "desedimentation." Husserl concludes:

> Existing in this way [as sedimented traditions], they [all the so-called exact sciences] extend enduringly through time, since all new acquisitions are in turn sedimented and become working materials. Everywhere the problems, the clarifying investigations, the insights of principle are *historical.*[7]

We must now turn our attention to these kinds of investigations, first in the history of mathematics and then in the history of mathematical physics.

Notes

1. See Heelan 1987.
2. Husserl 1970 [1939]. The essay was written in 1936 and first published in 1939, subsequent to Husserl's death in 1938.
3. Husserl 1978, 207–208.
4. Husserl did not himself coin or employ the term "desedimentation." He spoke instead of the "reactivation" of sedimented meanings and of the possibility for "cashing in" such meanings or recovering their original and intuitive self-evidence. Jacob Klein supplies a commentary on Husserl's reflections on history and the philosophy of science in Klein, "Phenomenology and the History of Science" (Klein 1985, 65–84). See also Hopkins 2011, chapter 1.
5. Husserl 1970 [1939], 354–355.
6. Husserl 1970 [1939], 363.
7. Husserl 1970 [1939], 369.

CHAPTER 4

The Historical Sense-Structure of Symbolic Algebra

Commenting in the appendix to his *Universal Arithmetic* on the use of Cartesian algebra in geometry, Isaac Newton expresses the following rather dim view of the practice:

> Equations are expressions of arithmetical computation, and properly have no place in geometry except so far as quantities truly geometrical (that is, lines, surfaces, solids, and proportions) may be said to be some equal to others. Multiplications, divisions, and such sorts of computations, are newly received into geometry, and that unwarily, and contrary to the first design of this science.... Therefore, these two sciences ought not to be confounded. The ancients did so industriously distinguish them from one another, that they never introduced arithmetical terms into geometry. And the moderns, by confounding both, have lost the simplicity in which all the elegancy of geometry consists.[1]

These strictures are in one sense surprising, coming as they do near the end of a treatise in which Newton subjects geometrical problems to extensive algebraic analysis. How do we explain them? In the first place, Newton is invoking the ancient distinction between continuous quantity (geometry) and discrete or numerical quantity (arithmetic). As is well known, after the discovery of incommensurability in the ancient Pythagorean School, Greek mathematics could no longer regard numerical methods in geometry as scientifically rigorous, given the problem of numerically undefined geometrical ratios. The situation was not substantively altered

J. K. Cosgrove, *Relativity without Spacetime*,
https://doi.org/10.1007/978-3-319-72631-1_4

by Eudoxus' general theory of proportion, which furnished a rigorous definition of geometrical ratios while leaving their numerical incommensurability unaffected. For Eudoxus did not discover the "real numbers," as is sometimes suggested.

We are used to representing the Pythagorean Theorem, for instance, as a sum of algebraic squares ($A^2 + B^2 = C^2$), the terms designating numbers multiplied by themselves. But consider Fig. 4.1 below from Euclid's *Elements* (Book I, Proposition 47).

Euclid's diagram depicts relations between *geometrical* squares built on the sides of a right triangle: squares *ABFG* and *ACKH* together add up to the square *BCED*. On this basis, were we to write the Pythagorean Theorem in the form of an equation it would be $ABFG + ACKH = BCED$, with no numbers or algebraic squares. The Pythagorean Theorem understood this way, in terms of a sum of geometrical squares, must not be regarded as the geometrical version of a properly algebraic theorem. Rather the reverse: The Pythagorean Theorem is about the relations between the geometrical squares; only subsequent to the proof based on geometrical squares can the theorem be translated numerically into algebra. The question raised by the passage from Newton above is whether the algebraic representation of the theorem sacrifices mathematical rigor by obscuring the geometrical relations involved. Euclid's proof of the Pythagorean

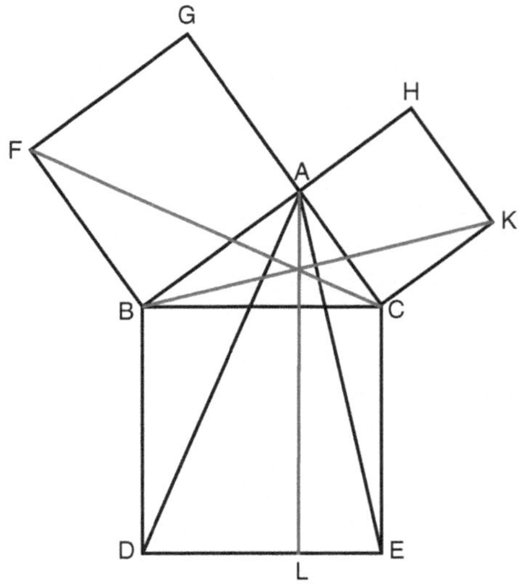

Fig. 4.1 Euclid's *Elements*, Book I, Proposition 47

Theorem does not treat the squares numerically, in the sense of relating their "areas" (length times width), for such treatment would require at least implicitly the designation of a numerical unit of length and therefore run up against the problem of incommensurability. So even if we write $ABFG + ACKH = BCED$, as above, we have an "equation" only in a notational sense, not in the sense of the moderns whom Newton chastises (chiefly Descartes, one must suppose). Unfortunately, these moderns have confused the subject matter of geometry by blurring the essential distinction between geometrical and numerical quantity.

But there is more at issue in the quoted passage than merely a nod on Newton's part to the superior rigor of the ancients in maintaining the distinction between continuous and discrete quantity. Newton singles out multiplication and division, "newly received" into geometry, for special censure, suggesting the employment of these operations in geometry has a more specific association with Cartesian algebra (the arithmetic of the moderns). Ancient Euclidean geometry freely employed addition, of course, although not in the numerical sense. Fortunately, an earlier discussion of multiplication in Newton's treatise sheds light on the present passage. In his initial discussion of the basic arithmetical operations Newton notes that, properly speaking, the multiplication of a line (by a number) yields a new line whose length is some numerical multiple of the original line. However, we also speak, albeit improperly, of "multiplying" one line by another line to yield a surface:

> Moreover, *custom has obtained, that the genesis or description of a surface, by a line moving at right angle upon another line, should be called multiplication of those two lines.* For though a line, however multiplied, cannot become a surface, and consequently this generation of a surface by lines is very different from multiplication, yet they agree in this, that the number of unities in either line, multiplied by the number of unities in the other, produces an abstracted number of [square] unities in the surface ....[2]

No doubt the ancient Greeks also knew how to multiply the numerical values of the sides of a rectangle to compute the area. However, they did not operate with "abstract numbers" in the modern sense of dimensionless entities susceptible to numerical computation. The abstract concept of "three," after all, cannot be multiplied by the abstract concept of "four" as if the abstractions were numbers in their own right. In this sense, our modern concept of number might be better termed "symbolic" than merely "abstract." In any event, Cartesian algebra employs multiplication and division routinely in Newton's "improper sense" above, presumably an additional lapse of rigor on the part of the moderns.

Clearly, Newton's strictures on algebraic equations in geometry apply, *mutatis mutandi*, to algebraic equations in physics. And, as we have observed already, the concept of Minkowski spacetime is essentially symbolic-algebraic in origin, notwithstanding its usual characterization as a departure from Einstein's algebraic methods. The founding analogy with the Pythagorean Theorem therefore remains open to question. To evaluate the analogy in terms of its physical validity, however, requires our first gaining a secure grasp of the algebraic constitution of concepts in modern mathematical physics generally. This task can be accomplished only through a philosophical analysis informed by historical scholarship. In the medieval science of mechanics, and up to the time of Galileo and beyond, relations between physical quantities were represented not numerically or algebraically, but rather geometrically in terms of ratio and proportion. It is not immediately clear how to translate a proportion of ratios involving homogeneous physical quantities into an equality of absolute numerical quantities while maintaining scientific rigor and physical intelligibility; not least because, as Newton saw so clearly, arithmetical operations in geometry and physics do not directly admit of such algebraic employment ("distance over time," for instance, or "mass times velocity," operations which were incoherent until the nineteenth century). Moreover, the concept of number itself undergoes a conceptual transformation in the sixteenth and seventeenth centuries coinciding with the rise of Cartesian algebra. The symbolic constitution of the modern number concept then carries over into the conceptual structure of mathematical physics itself, through the displacement of the traditional mathematics of ratio and proportion by modern symbolic algebra.

The conceptual obstacles to the assimilation of algebra into mathematical physics were nevertheless successfully overcome from the late seventeenth through the eighteenth century, eventuating not simply in a new mathematical language for the science of physics, but even more importantly a new conception of nature inseparable from symbolic mathematics.[3] We shall trace some relevant features of this development, first in the received Greek mathematical tradition of Euclid and Diophantus and then in the writings of the principal architects of modern symbolic algebra in the seventeenth-century: Vieta and Descartes. Fortunately, considerable scholarship on this development already exists, and I shall rely especially on Jacob Klein's classic study *Greek Mathematical Thought and the Origin of Algebra*.[4] While making no attempt to survey the history of algebra, I shall focus on specific developments relevant to the algebraic constitution of concepts in modern mathematical physics.

4.1 The Concept Number in Greek Mathematics

The concept of number (*arithmos*) in Greek mathematics, which governs the Western arithmetical tradition up to the sixteenth century, is set forth definitively in Book VII of Euclid: "A number is a multitude composed of units" (Definition 2).[5] It bears underlining that on Euclid's definition, a number is the multitude of units itself, not an abstract concept by means of which we refer to the collection of units. If the units in question are individual eggs, for example, then the number is those eggs themselves as a countable collection ("a dozen," as we say, but not "the number twelve"). Thus a number in the Greek conception is always determinate, both with respect to the kind of units (apples, eggs, and the like) and with respect to how many of them there are (*four* apples, a *dozen* eggs). Arithmetic, as a science of number, addresses pure or non-sensible units rather than sensible units like apples or eggs; but even in scientific arithmetic a number remains a multitude of (pure) units rather than an abstraction or symbolic entity.[6]

4.1.1 Arithmetical Operations in Euclid

The concept of number as a collection of countable units determines the arithmetical operations that can be performed on numbers. Thus we can add numbers together or subtract them from one another as long as the units are of the same kind. We cannot add apples and oranges, for instance, unless we change our unit of counting to *fruit*, thus restoring homogeneity of units. Multiplication raises additional considerations for ancient Greek arithmetic. Euclid defines the operation of multiplication with great care (*Elements*, Definition 15) in terms of *repeated addition*: "A number is said to *multiply* a number when the latter is added as many times as there are units in the former."[7] Note that in Euclid's definition the number being multiplied or the "multiplicand" is a collection of pure units, while the multiplier is a number of "takings" of that multiplicand. Therefore only the number being multiplied is an arithmetical quantity (collection of pure units) per se. That is, if I multiply four times three, for instance, I take *three pure units* four times. Commuting the operation by taking *four pure units* three times is, strictly speaking, a different operation.

In the Greek context, then, multiplication has a quite different meaning than our modern day sense of multiplying together dimensionless numbers. For on the Greek conception we do not really multiply a number by another number, but rather take a given number of pure units a certain number of times, by repeatedly adding the number to itself. The arithmetical science therefore employs what could be called a *natural* concept of number in the sense that the very same concept applies in the prescientific realm of sensible units. We could no more take three pure units four pure units times than we could take, for example, three eggs, two eggs times. Rather, we take a number of eggs, three of them, twice. The use of pure units in scientific arithmetic nevertheless lends itself to conceiving of those numbers in practice as if they were dimensionless, such that Euclid is indeed willing to speak of multiplying a number by another number. The early modern architects of our own science were well aware that the numerical methods of symbolic algebra (multiplying mass times velocity, for instance) raised issues of physical intelligibility and mathematical rigor in the science of physics.

Euclid understands the operation of division likewise in terms of one number being a "part" of another number or "measuring" it, as three measures six in the sense that six contains three twice. Division is thus the inverse of multiplication and should not be confused with fractions in the modern sense, by which, for example, $\frac{2}{4}$ can be understood both as "the number two divided by the number four" and as a fractional number in its own right. In the Greek context fractions necessitate a change of unit, since the unit itself, as the principle of number, cannot be fractionalized. Again, if I have three pizzas and cut each of them in half I can no longer count pizzas per se and must instead adopt as my new unit half a pizza. I now have six of these new units—six halves—and a whole pizza now comprises two units.

4.1.2 The Concept of Ratio

If a fraction in the sense of Greek mathematics should not be confused with the arithmetical operation of division, neither should it be confused with the quite different concept of a *ratio*. A fraction is a number of parts, while a ratio is a *relation* between numbers or more generally between quantities of the same kind. In the domain of number, of course, each particular number has the same ratio to the unit as it has units—the number three, for example, has the ratio 3:1 with the unit. But that does not render a ratio itself a number. By contrast, in modern day algebra ratios are generally represented as fractional numbers, effacing the very distinction between ratio (a relation between homogeneous quantities in general) and number (a countable quantity). Thus we represent the ratio between the circumference and diameter of a circle as the "dimensionless number" $\pi\left(\frac{C}{D}=3.14\right)$.

An operation on ratios of great importance for the historical transition to algebraic physics in the early modern period is the *compounding* of ratios. Euclid does not define compounding per se, but he does identify two specific types of compound ratio, the "duplicate" and the "triplicate" (Definitions 9 and 10 respectively). Given three magnitudes in continuous proportion (so $A : B : C$, with $A : B :: B : C$), the ratio $A : C$ "duplicates" $A : B$, and similarly for a triplicate ratio. In general, then, given two ratios $C : B$ and $B : A$, not necessarily in continuous proportion, the compounded ratio is $C : A$. A familiar shortcut for compounding numerical ratios is to multiply the antecedents and consequents and then place their respective products in ratio: (3 : 4)(4 : 5) :: (12 : 20). For this reason, the compounding of ratios is often defined as "multiplying" the ratios together, which translates into algebra as multiplication of fractions. That is, instead of compounding $a : b$ and $c : d$ to obtain the new ratio $e : f$, we write in algebra $\left(\frac{a}{b}\right)\left(\frac{c}{d}\right)=\frac{ac}{bd}$, which yields a number (quantity) rather than a ratio (relation). Strictly speaking, of course, a relation cannot be multiplied by another relation to yield a numerical quantity and such algebraically generated quantities are in reality symbolic artifacts.

These distinctions may strike the reader as somewhat naïve and in any case hardly relevant to modern day mathematical physics. Nevertheless, we prematurely formalize the operations of arithmetic at our peril, for any subsequently formalized sense they may acquire must be founded ultimately on their natural meaning in terms of the direct experience of countable quantities. That applies all the more with respect to the meaning of the formalized arithmetical operations in mathematical physics. For instance, do we genuinely grasp the physical meaning of the algebraically formalized operation "time squared" in $s=\frac{at^2}{2}$, the algebraic version of

Galileo's law of free fall? Does the algebraic quantity t^2 represent a physical quantity in nature? If it does not, what is its relation to the physical world? And what about mv or mc^2?

4.1.3 *Arithmetic and Geometry in Euclid*

Euclid never uses arithmetical operations in geometry, even if some of his arithmetical terminology has its origin in geometry. A "plane" number, for instance, which is obtained by multiplying two numbers ("sides"), is still a definite multitude of discrete units and can in no wise be assimilated to continuous geometrical quantity. In the arithmetical books of the *Elements* (VII, VIII, and IX), Euclid represents numbers in general, including plane and solid numbers, by simple line lengths, and there is no counterpart in Euclid to the later practice of "multiplying" a geometrical magnitude by another geometrical magnitude to obtain a magnitude of higher dimension. The use of geometrical terms such as square, cube, and so forth in Greek arithmetic no doubt has its origin in the arithmetical treatment of geometrical problems, a practice that must have been well established prior to the discovery of incommensurables, and which in any case would have been necessary for practical calculation. From a rigorous perspective, however, the use of arithmetical operations in geometry must be justified in terms of ratio and proportion, in particular the compounding of ratios. The size of a rectangle, for instance, increases with the compounded ratio of its sides (Euclid VI.23). Hence, if the lengths of the sides are increased by factors of a and b respectively, then the ratio of the new rectangle to the original rectangle is proportional to the ratio compounded of the ratios of the corresponding sides: $(a : 1)(b : 1) :: ab : 1$. If the factors are 2 and 3, for instance, the figure is now six times as large; and this result does not depend on our having first designated a unit of length and then multiplied the lengths of the sides by one another to obtain the area of the rectangle. We have obtained no numerical value for the absolute size of the resultant figure, that is to say, but instead determined the ratio of its size to the size of the original figure (6:1).

If, unlike Euclid, we desire a numerical treatment of the problem then we must first designate a unit length. Doubling and tripling the respective sides of a unit square now gives us a rectangle of *six square units*, since the consequent of our 6:1 ratio is now the unit square. And as long as the sides are numerically commensurable, as in the present case, we can take a shortcut by multiplying together their numerical values to obtain the "area." Nevertheless, in the Greek context we have not truly multiplied a "length by a length," for multiplication in the sense of a number of "takings" of a given length cannot yield a figure of higher dimension. If we multiply a length, we always simply obtain a longer length.

4.2 Algebraic Equations in Greek Mathematics: Diophantus of Alexandria

Modern symbolic algebra at its inception was a direct reinterpretation of the theory of equations in Diophantus.[8] While the influence of Diophantus' *Arithmetica* (third-century A.D.) was transmitted to the West indirectly from the thirteenth to the sixteenth-centuries through medieval Arabic algebra, it was nevertheless the direct appropriation of Diophantus that led to the algebra of Vieta and Descartes. Thus a review of Diophantine algebra is necessary to prepare us for consideration of those particular features of modern symbolic algebra that determine the concept of Minkowski spacetime.

4.2.1 The Concept of Number in Diophantus

The feature of Diophantine algebra that must be emphasized above all, in comparison to modern algebra from Vieta onward, is its consistency with the Euclidean definition of number. Diophantus gives the Euclidean formulation at the outset of his *Arithmetica*, Book I: "All numbers are made up of some multitude of units."[9] Nevertheless Diophantus calculates with fractions and negatives, even admitting the former—although not negatives—as solutions to problems. This practice has led some scholars to conclude that, notwithstanding his initial definition, Diophantus understands numbers in terms of what we call the *rational numbers.* According to Heath, for instance, Diophantus at least in his solutions countenances "no numbers whatever except 'rational' numbers; and in pursuance of this restriction he excludes not only surds and imaginary quantities, but also *negative* quantities."[10] Hence, for Heath, the Diophantine concept of number is effectively our concept of rational number minus the negative numbers. Diophantus thus abandons his original Euclidean definition of number as a multitude of units, something on which Heath is strangely silent. Another scholar goes further, maintaining that although Diophantus restricts solutions to positive numbers, he uses negative numbers in calculation, such that "it is safe to say that Diophantus extended the domain of numbers to the field of rationals ..."[11]

To determine the question, let us first consider how Diophantus presents algebraic expressions in general. A typical problem in the *Arithmetica* is to find some number or numbers with stipulated relations to other numbers, which as a rule entails an initial assignment of determinate values to given numbers. In the very first problem of the *Arithmetica*, for instance, Diophantus seeks to "divide a given number into two having a given difference."[12] The given number is then stipulated as one-hundred and the difference forty. With the smaller number designated as the unknown, Diophantus in our notation constructs the equation $2x + 40 = 100$. All his symbols are abbreviations for words and Diophantus' symbolism here says, "Two times the unknown number plus forty units equals one-hundred units."

While Diophantus does lay down rules for multiplying negatives, this does not signal the concept of a "negative number," but rather simply facilitates calculations such as the reduction of $\frac{a}{3} = 2x - 10$ to $a = 6x - 30$ (I.21). In the reduced expression, that is, we have multiplied the number 10 by the number 3, yielding the number 30 to be *taken away* from $6x$; we have not multiplied the number "-10" by the number 3 to yield the number -30. Hence, when subtraction is involved Diophantus takes a (positive) number of units away from some other number of units, and if the result would go less than zero he rejects it. Cleary if Diophantus truly countenanced negative numbers he would have no reason to preclude them in his solutions.

What about fractions? To be sure, Diophantus employs them freely both in calculation and in some solutions (for example, I.21, I.23, I.24, I.25, and I.39). On the other hand, when the conditions of the problem permit he converts fractional solutions into "whole number" solutions by multiplying through by a common denominator, a practice that calls to mind Plato's remark in *Republic* VI (525E) that if someone proposed to partition the unit or monad, a trained mathematician would laugh at him and proceed to multiply. In such cases Diophantus has found it necessary to employ an auxiliary unit in the course of the calculation; then, upon obtaining a fractional solution, he adopts that auxiliary unit as the new unit of the problem. In the course of calculation in I.24, for example, he obtains $x + \left(x + \frac{1}{3}\right) + \left(x + \frac{1}{2}\right) = x + 3$, yielding $x = \frac{13}{12}$ for the solution of the unknown along with $\frac{17}{12}$ and $\frac{19}{12}$ for the other two numbers sought by the problem. Here $\frac{1}{3}$ means *one of three fractional parts into which the original unit has been* partitioned—it is not the "rational number" $\frac{1}{3}$ in our sense, but rather exactly *one* of the auxiliary units (those new units being *one-third parts* of the original unit).

Hence, for Diophantus, just as for Greek arithmetic in general, the unit qua unit is indivisible, even though it can be replaced with a new unit to accommodate the exigencies of calculation. Diophantus procedure is akin, for instance, to what we are used to doing with denominations of money. In American currency the dollar is the unit, but we still count pennies or "cents" ($\frac{1}{100}$ of a unit) as if they were units in their own right (auxiliary units, as I have termed them above). In Diophantus' case, the mere employment of an auxiliary or provisional unit does not change the defined units of the problem (just as a penny is still a cent or $\frac{1}{100}$ of the American monetary unit). In the problem under consideration above (I.24), however, Diophantus' last step is to change the unit of the problem by multiplying through by the common denominator twelve to obtain the solution 13, 17, and 19 *in the new units*. The provisional unit has become the unit of the problem.

The Diophantine monad or "one" therefore exhibits, in its practical divisibility, all the characteristics of a sensible unit of measure (meter, second, or the like), while maintaining the status of a pure unit, indivisible per se and not tied to any particular kind of sensible matter. A more formidable challenge to the thesis that Diophantus affirms in practice the Euclidean definition of number as a determinate multitude of units is his initial definition of the unknown as an "indeterminate multitude of units" (*plēthos monadōn aoriston*)[13] as well as several lemmas (IV.34, 35, and 36) in which he proposes to find "indeterminate numbers" (*arithmous aoristous*).[14] Such phraseology might seem to suggest that a number for Diophantus is not necessarily a determinate number of units but rather some kind of more generalized entity, which Diophantus for some reason feels the need to render determinate in the solutions to his problems. This is evidently how Vieta in the sixteenth century took Diophantus, but as Klein points out, such an interpretation betrays the dubious assumption that Diophantus must be "groping" for the full generality of our modern algebraic concept of number. An indeterminate number in the Diophantine context, however, means simply a definite number that has *yet to be determined*, not an indeterminate "generalized quantity" in the modern sense.[15] If I have some eggs in the refrigerator but do not know how many there are, I might refer to there being an "indeterminate number" of eggs in the refrigerator; but in fact there are just so many eggs in the refrigerator and I intend nothing else than that by calling the number indeterminate. That is, the number of eggs is *undetermined* but not indeterminate. Accordingly, those lemmas where Diophantus seeks indeterminate solutions are always auxiliary to the following problem, where the numbers left undetermined by the previous lemma take on determinate values. In these particular problems and in other indeterminate ones such as I.22 through I.25, Diophantus is therefore supplying a general procedure for obtaining a determinate result. As Klein observes, with Diophantus "we must distinguish strictly between the *procedure* and the *object*; while the procedure is … 'general' (*katholou*), the object intended is in each case a determinate number of monads [units]."[16]

4.2.2 Algebraic Calculation with "Species"

Diophantus' calculation with what he calls "species" (*eidē*) must be understood in terms of his understanding of number, discussed above, as a *determinate* multitude, even when that multitude is provisionally undetermined. Diophantus characteristically notes that the very definition of number as a multitude of units implies that number can always be augmented to infinity by the successive addition of units. Thus a scientific knowledge of number requires that the numbers be classified into *eidē* or kinds, for otherwise the trained mathematician could not "know all the numbers," as Plato observes in *Theatetus* 198A-B. For Diophantus, such classification of numbers into *eidē* or species is concerned, in the first place, with the classifications for the "square" (x^2), the "cube" (x^3), the "square-square" (x^4), and so forth, but also the unknown itself as well as the given number or "constant" in our sense (which Diophantus however always assigns a definite value). Thus Diophantine species are not themselves numbers, but rather classes of numbers. The square numbers, for instance, are 4, 9, 16, 25, and so forth, designated indeterminately for purposes of calculation since they are not yet known. Accordingly, the sign for the square (Δ^Y) intends not an indeterminate quantity but rather a determinate quantity as yet unknown, to be discovered in the course of solving the problem. For this reason Diophantus always gives his solutions in terms of determinate numbers, never generalized solutions (save for the provisional cases discussed above).

A number of algebraic manipulations that we take for granted today are precluded by Diophantus' number concept. In the first place, there can be no negative, irrational, or imaginary numbers. Nor can there be continuous functions of the form $f(x) = y$, with associated "graphing" of equations based on the continuity of the real number line. Certainly there can be no multiplication of quantities by one another, such as mass times velocity, nor division of heterogeneous quantities (distance over time, for instance). If algebra is to be of use in mathematical physics, then, the number concept of Diophantus must undergo the requisite alterations, which it does at the hands of Vieta and Descartes.

4.3 Modern Symbolic Algebra

François Viète (Vieta) (1540–1603) is generally recognized as the founder of modern symbolic algebra. While Vieta's "general art" is anticipated in certain respects by medieval Arabic algebra, modern symbolic algebra takes its point of departure from Vieta's direct reinterpretation of the Diophantine concept of "species." This is the algebra, with improvements by Descartes and others, assimilated into European mathematical physics beginning around the late seventeenth century.

Vieta sketches the basic principles of what he calls symbolic logistic or "species calculus" (*logistice speciosa*) in *Introduction to the Analytical Art* (*In Artem Analyticem Isagoge*) of 1591. It is clear from the outset of this document that Vieta conceives of a general mathematical art encompassing quantity of any kind, be it geometrical or numerical: "Our art," he writes in his opening dedication to his patroness Catherine of Parthenay, "*is the surest finder of all things mathematical.*"[17] However, while the principal ancient model for such a universal mathematics, Eudoxus' general theory of proportion (Euclid, Book V), treats both geometrical and numerical subject matter through a general method, while maintaining the essential distinction between the two kinds of quantity, Vieta generalizes the *object* of mathematical analysis in such a way that the fundamental Greek distinction between continuous and discrete quantity is effectively obliterated. Indeed, according to Jacob Klein, whose interpretation we shall endorse here, Vieta introduces a new generalized or symbolic conception of number that governs the subsequent development of European mathematics. Thus Vietan algebra evinces a fundamental discontinuity with ancient Greek mathematics, while at the same time interpreting itself as a direct development of that ancient mathematics. The decisive step is Vieta's symbolic reinterpretation of the Diophantine concept of "species."

4.3.1 Vieta's Reinterpretation of Diophantine Species

The Diophantine *eidos* ("species"), we recall, designated by a letter sign, is not itself a number but rather the kind or class to which an intended but as yet undetermined number belongs (square, cube, and so forth, or simply the unknown itself). Hence, although Diophantus does subject the letter sign directly to arithmetical calculation, it is not the species per se that is the intended object of calculation, but rather a determinate number (of units) falling under that particular type. A Diophantine equation we would write down as $2x^2 = 50$, for instance, means that the number we are looking for is a square number which, taken twice, makes fifty units. Along the same line of thinking, when Diophantus affirms that any species multiplied by unity remains the same (Definition Six), he does not mean that any number multiplied by one equals itself (our "multiplicative property of 1"), but rather that any number taken once remains in the same class or species as before—if it was a square number it remains a square number and so forth.[18] By contrast, Vieta's analytical art or "zetetics" reinterprets the Diophantine species, designated by a letter sign, as "magnitude in general" while *at the same time* subjecting it to arithmetical calculations as if it were itself a number. Thus Vieta prefaces his list of rules for symbolic calculation with the remark that "[t]he numerical reckoning operates with numbers; the reckoning by species operates with species or forms of things ..."[19] By means of the transformation of the Diophantine species concept, "number" is reconceived in Vieta's analytical art as indeterminate or generalized quantity, no longer a "number of" countable units per se. This development expresses itself not only in how Vieta contrasts his analysis with that of Diophantus but above all in how he understands what he calls the "law of homogeneity."

In Chapter 5 of the *Analytical Art*, Vieta praises Diophantus for the generality of his algebraic method, deliberately concealed:

> Diophantus in those books which concern arithmetic employed zetetics most subtly of all. But he presented it as if established by means of numbers and not also by species (which, nevertheless, he used), in order that his subtlety and skill might be the more admired; inasmuch as those things that seem more subtle and more hidden to him that uses the reckoning by numbers (*logistice numerosa*) are quite common and immediately obvious to him who uses the reckoning by species (*logistice speciose*).[20]

Whereas in reality Diophantine analysis, however general in procedure, is always ruled by the understanding of number as a determinate multitude of countable units, Vieta interprets Diophantus as having attained his results by means of a calculus of generalized quantity, as Vieta himself. Likewise in Chapter 1 Vieta stresses that, unlike the analysis of the ancients, his new zetetic art "does not employ its logic on numbers—which was the tediousness of the ancient analysts—but uses its logic through a logistic which in a new way has to do with species."[21]

What is especially significant in these passages is that Vieta conceives his symbolic logistic as numerical in character and therefore in continuity with Diophantine analysis, even though the Vietan calculus proceeds with "species" understood as generalized symbolic magnitudes.[22] The kind of generality at issue, therefore, is not merely the generality of a procedure equally and indifferently applicable to both geometrical and numerical quantity; for even if generality in that sense is one of Vieta's principal desiderata there would be no point in mentioning Diophantus as a predecessor in this respect. Rather, Vieta's innovation is a generality with respect to the *object* of calculation and solution, which is solely *indeterminate quantity*, conceived *numerically* via the arithmetical calculus of letter symbols.

4.3.2 Vieta's "Law of Homogeneity" and the Symbolic Concept of Number

The new number concept with which Vieta is operating comes through most clearly in his novel formulation of the "law of homogeneity" (*Introduction*, Chapter 3), which is "[t]he supreme and everlasting law of equations and proportions ...."[23] The principle of homogeneity in Greek mathematics, whose authority Vieta indeed cites in the text under consideration, specifically regards ratios. Euclid defines a ratio [*logos*] as "a sort of relation [*skēsis*] in respect of size between two magnitudes of the same kind" (Euclid, Book V, Definition 3).[24] There exists no ratio, for instance, between a plane and a line. By contrast, Vieta's law of homogeneity regards not ratios but the numerical operations of addition and subtraction:

> The supreme and everlasting law of equations or proportions, which is called the law of homogeneity because it is conceived with respect to homogeneous magnitudes, is this: Only homogeneous magnitudes are to be compared with one another.... And so, if a magnitude is added to a magnitude, it is homogeneous with it.[25]

So only a magnitude of the same kind can be added to or subtracted from another magnitude. That might seem to go without saying, but in terms of Vieta's calculus of species the requirement means specifically that every term of an algebraic equation must be of the same degree or dimension. A plane magnitude can only be added to another plane magnitude, for instance, and so an equation such as $x^3 - ay = y^3$ (in Vieta's notation, *A cube minus B plane in E equals E cube*) satisfies Vieta's law of homogeneity only if *a* is understood as a so-called "plane magnitude."

The homogeneity requirement applies solely to addition and subtraction, though, whereas multiplication in Vieta's calculus yields a heterogeneous magnitude (a plane times a side yields a solid, for instance).[26] Evidently, while the basic arithmetical operations (addition, subtraction, multiplication, division) get their sense from the ordinary arithmetical operations of the same name, the numerical meaning of these ordinary operations has been altered in a certain way. In Euclid or Diophantus, the geometrical terminology "side" or "plane" as applied to numbers is simply a classification and so there is no more of an impediment to adding a side to a plane than there is, for instance, to adding an even number to an odd number. Similarly, multiplication of a side and a plane yields a new classification (cube), but the number produced is still a multitude of countable units, just as if one were to multiply an odd number by an even number to yield a number of a different class (even). But in Vieta's calculus not only must addition and subtraction be limited to terms of the same dimension, but both multiplication and division yield quantities of different dimension. Clearly we are no longer dealing with a "number of" something (apples, dogs, pure arithmetical units).

Witmer observes in the introduction to his translation of Vieta that "[n]ot everything … Viète proposed has served the test of time. One item in particular is his insistence on endowing coefficients with dimensions such that all terms in any given equation will be of the same degree."[27] Indeed, any satisfactory interpretation of Vieta's law of homogeneity must account for its precipitous abandonment by his successors. Already in Descartes' *Geometry* Vieta's law of homogeneity is reduced the requirement that terms of heterogeneous degree be either multiplied or divided by unity until they are of the same degree. The usual explanation for the eclipse of Vieta's law of homogeneity is that Vietan magnitudes are themselves geometrical or at least Vieta is constrained by a traditional way of geometrical thinking in algebra. Morris Kline, for instance, remarks that, "Vieta too was largely tied to geometry. Thus he writes $A^3 + 3B^2A = Z^3$, where A is the unknown and *B* and *Z* are constants, in order that each term be of third degree and so represent a volume."[28] As Vieta stresses, however, the proximate object of zetetics is magnitude in general ("forms of things"), not geometrical magnitude per se. Only at the stage of "exegetics," subsequent to having solved the equation in general terms by means of zetetics or symbolic logistic, does the geometer or arithmetician as the case may be assign a specifically geometrical or arithmetical sense to the species upon which the zetetic art operates.[29] Moreover, Vieta explicitly maintains that equations are "resolutions" of proportions in general and so he uses powers higher than the third as a matter of course.[30] If $y : x^2 :: x^2 : a$, for instance, then $ay = x^4$. As for the notion of a traditional limitation on terms of different degree in the same equation, there in fact was no such limitation. Diophantus adds and subtracts terms of different degree without comment, since, as we already observed, his square, cube, square-square and so forth are understood alike as numbers of countable units. A geometrical requirement of homogeneity could arise for Vieta only if he were employing geometry as a method of proof in algebraic problems. Cardano in his *Ars Magna* of 1545, for instance, solves the cubic equation $x^3 + 6x = 20$ geometrically,[31] and clearly if the numerical quantity x^3 is to be represented for purposes of proof by a geometrical cube, then $6x$ must be a solid and, accordingly, 6 a plane. But there is no homogeneity requirement imposed on the equation itself, only on the geometrical proof. Vieta's law of homogeneity, by contrast, is numerical in origin.

If we remind ourselves that Vieta's homogeneity requirement applies specifically to *species* or generalized magnitudes, not to collections of countable units per se, we can see that it is solely with reference to the new indeterminate or generalized number concept that Vieta must stress his homogeneity requirement—for the numerical operations of addition and subtraction are intelligible only if carried out with reference to some unit of calculation. For Vieta that unit (the "homogeneous element"[32]) is now "one plane unit" or "one solid unit" and so forth, or simply one "side" without any adjoined power designation at all. Such a unit or homogeneous element is not geometrical per se, even if it obviously depends on a geometrical analogy: for it is simply a unit of general magnitude, conceived numerically but indifferently applicable to geometry or arithmetic. Just as clearly, Vieta's unit is no longer arithmetical in the sense of Euclid or Diophantus, for when number is understood in the traditional sense as a "number of" countable units, homogeneity need hardly be emphasized since it is fulfilled as a matter of course. On this point Jacob Klein is worth quoting at length:

> Vieta's law of homogeneity is concerned … with the fundamental fact that every "calculation," since it does, after all, ultimately depend on "counting off" the basic units, presupposes a field of *homogeneous monads.* For Diophantine "logistic" this demand is fulfilled as a matter of course, because it already operates within such a field of "pure" monads—the known and unknown "magnitudes" which are there united in an equation represent, each and every one, an "*arithmos* of monads." For the *logistice speciose* this fundamental presupposition needs to be especially stressed; hence the emphasis with which Vieta, in contrast to the "ancient analysts" (veteres Analystae), expounds his "lex homogeneorum" as the foundation of the "analytical art." Thus it appears that the concept of the species is for Vieta, its universality notwithstanding, irrevocably dependent on the concept of "*arithmos.*" *He preserves the character of the* "arithmos" as a "*number of …*" *in a peculiarly transformed manner.* While every *arithmos* intends *immediately the things or the units themselves* whose number it happens to be, his letter sign intends *directly the general character of being a number* which belongs to every possible number, that is to say, it intends "number in general" immediately, but the things or units which are at hand in each number only mediately.[33]

From our vantage point Vieta's law of homogeneity may seem like an empty gesture, a geometrical anachronism intruding upon the dimensionless realm of pure number. But our perspective is conditioned by the historically sedimented logic of the Vietan conception of number itself, in which number is no longer directly determined by reference to a field of countable units.

Klein proceeds in the same discussion to characterize this conceptual transformation in terms of the scholastic distinction between a "first intention" (*intention prima*) or direct concept of a being and a "second intention" (*intention segunda*) or concept which directly intends another concept rather than a being. In Greek mathematics the concept of number is first intentional or directly "about" some determinate collection of countable units. Vieta takes the indeterminate "species" or concept of a number *in general* (second intention) as if it were the object of a first intention by subjecting it to numerical calculations, a conceptual innovation rendered possible by the designation of the "homogeneous element." Klein continues:

> Furthermore—and this is the truly decisive turn—this general character of number or, what amounts to the same thing, this general number in all its indeterminateness, that is, its merely possible determinateness, is accorded a certain independence which permits it to be the subject of 'calculational' operations.[34]

Through the rules of the symbolic calculus, then, the mathematical object intended by Vieta's algebraic letter sign has essentially merged in meaning with the letter sign itself; and this is what Klein refers to as the modern *symbolic conception of number*, in contradistinction to the mere use of symbols to represent numbers. In Vieta's analytical art, that is, the means of representation in a sense has been identified with the thing represented. This is most evident for species of the first degree, since in such case one calculates simply on a letter sign itself, with no adjoined power designation and therefore no evident numerical character to the sign other than that which accrues to it via the operational rules for addition, subtraction, multiplication, and division (set forth by Vieta in Chapter 4 of his *Analytical Art*). Moreover, the symbolic mode of conceiving number rapidly comes to govern not merely algebraic variables like x or y, but also determinate numbers in our sense like "the number 4." The common observation that Greek mathematics recognized solely the "natural numbers," as if the Greek domain of number were defined simply in terms of some subset of our own, thus betrays a failure of historical perspective.

4.3.3 Vietan Algebra as Mathesis Universalis

We can now more precisely characterize the difference, with regard to the kind of generality at issue, between the ancient general theory of proportion and modern "general algebra" or *mathesis universalis*. Eudoxus' general theory of proportion, as set forth in Euclid, uses line lengths to symbolize proportions in general, whether geometrical or numerical. But the ancient theory does not identify this particular means of representation, the symbolic line lengths, with the subject matter under study. Hence, the general theory of proportion maintains the distinction between numerical and geometrical ratios. By contrast, modern symbolic algebra essentially identifies the symbolic means of representation—the letter sign or "species"—with the thing represented ("magnitude in general"), by subjecting the letter signs to numerical operations as defined by the rules of the symbolic calculus. Since the concept of magnitude in general makes no distinction per se between discrete and continuous quantity, that distinction is lost to modern symbolic algebra.

This very development is evident already, in fact, even before Vieta, in Simon Stevin's *Arithmetique* of 1585.[35] Stevin defines number not as a multitude of countable units but rather as "that by which the quantity of each thing is understood" ("*cela, par lequel s'explique la quantité de chacune chose*"),[36] explicitly rejecting the connection between number and discrete quantity per se.[37] For according to Stevin, and contrary to the traditional understanding, the unit itself is a number.[38] Even today in everyday speech I might say, for instance, "I have a number of errands to run," meaning at least two of them. Stevin however appeals to the principle that the part is of the same material as the whole to argue that the unit, always being a part of other numbers, must itself be a number. Just as slicing up a loaf of bread yields smaller pieces of bread, so also slicing up number in any way yields simply number. The conclusion follows only if we implicitly understand the concept of number in the symbolic sense identified by Klein, though, since only on that account is the "material" of number of the same nature as number: That is, the general concept of *being a number* is now understood as the material of number rather than the material of a number being the discrete units which make up a number.[39] Hence, based on his implicitly symbolic conception of number, Stevin can introduce "irrational numbers" in our modern sense and in principle the real number line itself. The symbolic number concept thus removes one of the traditional barriers to a numerical and, *mutatis mutandi*, algebraic treatment of continuous quantities in mathematical physics (distance, time, mass, and so forth). With respect to mathematical physics, however, an even more decisive feature of Vieta's analytical art is its explicit treatment of equations in general as symbolic translations ("resolutions," as he calls them) of proportions. For example, in Chapter 2 of the *Analytical Art*, after setting forth the "stipulations governing equations and proportions," which, he maintains, are already to be found in Euclid, Vieta observes that in view especially of his stipulations #15 and #16 on the multiplication of extreme terms (if two ratios $a : b$ and $c : d$ are proportional, for instance, then $ad = bc$), a proportion can be called the "composition" (*constitutio*) of an equation and an equation the "resolution" (*resolutio*) of a proportion.[40] No such identification of proportions and equations is possible for Diophantus, since numerical proportions comprise a mere subset of proportions in general. The Vietan-Cartesian understanding of equations as symbolic-analytic translations of proportions will subsequently determine the algebraic structure of concepts in modern mathematical physics.

4.4 Descartes and Symbolic Space

Descartes is clear from the beginning that algebraic equations are symbolic translations of proportions, explicitly stressing the relationship between the two in his *Discourse on Method* of 1637, to which Descartes' treatise on algebra (the *Geometry*) is appended as a demonstration of the scientific method therein articulated:

> For I saw that, despite the diversity of their objects, they [the mathematical sciences] all agree in considering nothing but the diverse relations or proportions that hold between these objects. And so I thought that it best to examine only such proportions in general, supposing them to hold only between such items as would help me to know them more easily.... Next I observed that in order to know these proportions ... I should suppose them to hold between lines, because I did not find anything simpler, nor anything that I could represent more distinctly to my imagination and senses. But in order to keep them in mind or understand several together, I thought it necessary to designate them by the briefest possible symbols. In this way I would take over all that is best in geometrical analysis and in algebra, using the one to correct all the defects of the other.[41]

For Descartes, therefore, equations are symbolic translations of proportions involving general magnitudes, with such magnitudes visually represented by line lengths. The symbolic conception of line lengths as general magnitudes is Descartes' particular innovation—not merely the symbolic employment of line lengths to represent numbers or other quantities, which of course can be found even in Euclid—but the symbolic *conception* of line lengths as general magnitudes, in the same sense as the symbolic number concept Descartes inherits from Vieta.

4.4.1 Geometrical Representation of Arithmetical Operations

Descartes famously announces at the very outset of his *Geometry* that "[a]ll the problems of geometry can easily be reduced to such terms that thereafter we need to know only the length of certain straight lines in order to construct them."[42] Forthwith he sets forth his method for carrying out the basic operations of arithmetic geometrically by means of the theory of ratio and proportion. To multiply two lines together, for instance, we place one of them (BD) in ratio with unity ($AB = 1$) and the other (BC) in ratio with the desired product (BE). We then construct similar triangles as in Fig. 4.2 below:

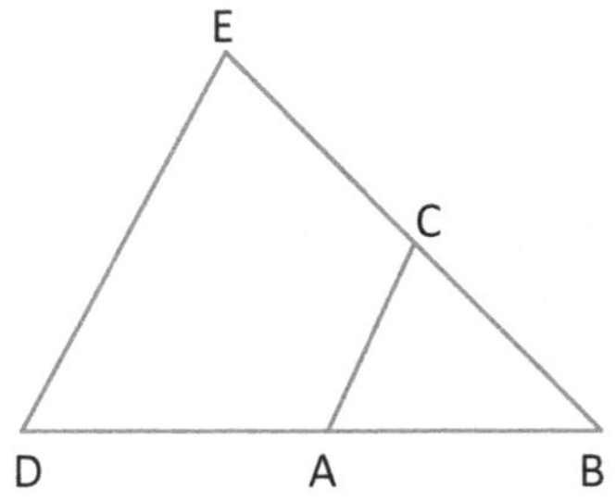

Fig. 4.2 Multiplication in Descartes' *Geometry*

It follows by the theorem of similar triangles that *BD* : 1 :: *BE* : *BC* and so (*BD*)(*BC*) = *BE*. Note that the multiplication of line by a line now yields another line rather than a plane, and so Descartes is no longer limited to three dimensions in his geometrical representation of algebraic quantities. "In one brilliant insight," Grosholz observes, Descartes "has freed the algebra of magnitude from the constraints and complications that hampered Viéte."[43] Of course, Descartes entitles his book *Geometry* rather than *Universal Arithmetic* because his ultimate aim is to represent geometry algebraically; and this aim will once again raise the issue of the numerical unit.

Having introduced his geometrical arithmetic Descartes turns to geometry proper, remarking disapprovingly upon the "scruple which the ancients had [against] using the terms of arithmetic in geometry, which could proceed only from the fact that they did not see clearly enough their relationship …" ("*le scrupule, que faisoient les anciens d'user des termes de l'Arithmetiqueen la Geometrie, qui ne pouuoit proceder, que de ce qu'ils ne voyoient pas assés clairement leur rapport*").[44] For example, the ancient geometer Pappus assumed that the famous locus problem is limited to six lines, since there can be no figure of higher than three dimensions.[45] The limitation to six can be removed, according to Pappus, only by conceiving the problem in terms of compounded ratios among the given lines. But for Descartes, Pappus' way around the limitation betrays a failure at grasping the true relationship between geometry and arithmetic, which is numerical. That is, the compounded ratios proposed by Pappus should properly be expressed *numerically* by means of an equation, in terms of the arithmetical operation of multiplication. And when we multiply the lines to obtain new lines, clearly there is no limitation on how many lines may be so multiplied. Consequently, Descartes easily solves the locus problem by means of his algebraic calculus or, as we call it, "analytical geometry."

The resolution of a proportion into an equation clearly is a powerful mathematical tool, and it would be just a short step, which Descartes himself nevertheless does not explicitly take, to dispense altogether with the cross-multiplication of proportions and simply nominalize ratios directly as fractional numbers. The nominalization of ratios as numbers is indeed suggested by the very fact that every number is the antecedent of its own ratio with the unit (e.g., $\frac{3}{1}:1::3:1$).[46] Pursuing this line of thought, John Wallis in his *Mathesis Universalis* of 1657 defines numbers themselves as simply "indices" of all possible ratios with the unit,[47] essentially the definition of number given in the twentieth century by the mathematician Hermann Weyl: "Numbers are merely concise *symbols* [my italics] for such relations [ratios] …"[48] Since a ratio, being a relation rather than a quantity, has no dimensionality per se, numbers themselves are rendered dimensionless. The equations of modern mathematical physics would be impossible without the nominalizing of ratios as dimensionless numbers, a historically sedimented development essential to the meaning-structure of the concepts of modern physics, as when we translate Galileo's law of fall as $s=\frac{at^2}{2}$.

4.4.2 Descartes' Symbolic Interpretation of Geometrical Magnitude

In Descartes' algebra—and this is what we especially must underline—*magnitude in general*, the proximate object of the Cartesian *mathesis universalis*, is to be treated numerically regardless of the ultimate subject matter under investigation. Indeed, geometrical magnitudes function in Descartes' *Geometry* at two levels simultaneously: in the first place as symbolic representations of generalized magnitude and in the second as the geometrical subject matter under study. The subject matter under study, therefore, could always have been otherwise: "At the same time I would not restrict them [proportions] to this alone [geometrical magnitudes], in order that I could better apply them afterwards to all the others to which they might be suited."[49] The same holds in Rule Fourteen of the early *Rules for the Direction of the Mind*, where Descartes stresses that "when the terms of a problem have been abstracted from every subject … then we understand that all we have to deal with here are magnitudes in general."[50]

Given Descartes' aim, which is the numerical-algebraic treatment of generalized magnitude, it is no surprise when in Book One of *Geometry* he announces his own law of homogeneity, such as we have already encountered in Vieta:

> It is also to be noted that each of the parts of a single line should ordinarily be expressed by as many dimensions as each other part, when unity is not determined in the problem: thus, here, a^3 contains as many dimensions as abb or b^3, which compose the line which I have called $\sqrt[3]{a^3 - b^3 + ab^2}$. But it is not the same thing when the unity is determined, because unity can be understood throughout, [even] where there are too many or too few dimensions; thus, if it is necessary to extract the cube root of $a^2b^2 - b$, we must consider that the quantity a^2b^2 is divided once by unity, and that the quantity b multiplied twice by the same.[51]

But from where precisely does the need for dimensional homogeneity arise? Descartes is clear that all the terms in his equations are simply lines, so in the passage above he cannot be referring to homogeneity of *geometrical* dimension; as he says, he uses such geometrical terms solely as a concession to standard terminology in algebra ("Here it is to be noted that by a^2 or b^3 or the like, I ordinarily mean only simple lines, although, in order to make use of the names used in algebra, I call them squares, cubes, etc."[52]) Rather, since Descartes' algebraic letter signs, like Vieta's, designate "general magnitude" while at the same time maintaining a specifically *numerical* intelligibility, some unit of calculation must be defined. To convince ourselves that Descartes' letter signs do indeed carry a numerical intelligibility, we need only underline the interpretation of algebraic terms such as a^2 or b^3 above as "simple lines," which would be unintelligible were Descartes' calculus a purely geometrical algebra. Descartes overcomes Vieta's proscription on terms of unlike degree by designating a unit line length and multiplying or dividing by it to obtain the required homogeneous unit of numerical calculation. Thus it is not the case, as is often suggested, that Descartes here overcomes a "traditional impediment" in algebra, tied to the geometrical conception of quantity; for as we saw with Vieta there could be no such impediment in algebra itself. Rather, it is the Vietan symbolic concept of number that first gives rise to the issue of homogeneity, for only when numerical calculations, which do presuppose a field of homogeneous units, are performed upon symbolically conceived letter signs does the status of the numerical unit become an issue.

It is, therefore, no traditional impediment but rather this very modern impediment that Descartes removes through multiplication by the unit.[53]

4.4.3 *Symbolic Space*

Although Descartes in his *Geometry* does not actually employ "Cartesian coordinates" in our sense of perpendicular axes with numerical coordinates assigned to points, in principle the method is nevertheless in play. In his solution to the locus problem, for instance, Descartes refers all quantities to two designated reference line segments (x and y) and determines the desired locus of points by means of numerical distance and angle to the reference segments. Clearly, every point by this means can be fixed in location with reference to x and y, so we effectively have a coordinatized space. However, since Descartes' line segments are symbolic magnitudes that can represent quantity of any kind, geometrical or not ("all the others to which they might be suited"), Descartes has in essence set up a *symbolic space* (later called a configuration space) within which quantitative relations in general can be represented. We call such representations "graphs" of equations.

An ellipse in Cartesian symbolic space, then, is proximately the graph of the equation for an ellipse, which may or may not have anything to do with a geometrical ellipse. Yet in *Geometry* Descartes betrays little awareness that he is operating simultaneously at two different levels of representation: the symbolic ellipse, representing an equation, on the one hand, and the geometrical ellipse itself on the other. As Klein points out, Descartes rather seems to assume that his symbolic figures are the same ones the ancients drew to directly represent or "image" figures of interest to the science of geometry.[54] Even in geometry per se, then, a Cartesian symbolic ellipse directly represents not a determinate geometrical figure, but rather the general character of being an ellipse (the general quantitative relations that define "ellipse as such").

On this basis Klein sees in Cartesian "symbolic extension" a transformation of the object of a second intention (symbolic ellipse) into a first intention (geometrical figure), analogous to the construction of Vieta's symbolic number concept. That is, the symbolic ellipse or visual representation of the equation for an ellipse in general (we write $\frac{x^2}{a^2}+\frac{y^2}{b^2}=1$) Descartes interprets as a geometrical figure in its own right; and, more generally, since Descartes identifies matter with geometrical extension ("my entire physics is nothing but geometry"[55]), Descartes has effectively cast his symbolic space or "Cartesian coordinate space" as the substance of the physical world. Indeed, Klein concludes his analysis of Descartes' algebra with the pregnant observation that Cartesian symbolic space subsequently becomes the "absolute space" of Newtonian physics: "Only at this point has the conceptual basis of 'classical' physics, which has since been called 'Euclidean space,' been created. This is the foundation on which Newton will raise the structure of his new science of nature."[56]

Klein's interpretation of Cartesian symbolic space is confirmed in a certain way by Einstein's own treatment of the concept of space in two of his essays from the 1930s ("The Problem of Space, Ether, and the Field in Physics," 1934[57]; and "Physics and Reality," 1936[58]). Einstein proposes that the intuitive origin of the concept of space lies in our direct experience of solid bodies. If two solid bodies are separated by an interval we can interpose a third solid body that exactly fills that interval and, assuming the rigidity of the third body, exactly fills any other equal interval. Thus we have the concept of a measurable interval in space. But this is not yet the concept of a mathematical continuum of space, which, according to Einstein, we first encounter in Descartes' method of numerical coordinates:

> In the geometry of the Greeks space plays only a qualitative role, since the position of bodies in relation to space is considered as given, it is true, but is not expressed by means of numbers. Descartes was the first to introduce this method.[59]
>
> Space as a continuum does not figure in the conceptual system [of Greek geometry] at all. This concept was first introduced by Descartes, when he described the point in space by its coordinates. Here for the first time geometrical figures appear, in a way, as parts of infinite space, which is conceived as a three-dimensional continuum.[60]

Einstein then concludes his analysis on a note strongly suggestive of Klein:

> In so far as geometry is conceived as the science of laws governing the mutual spatial relations of practically rigid bodies, it is to be regarded as the oldest branch of physics. This science was able, as I have already observed, to get along without the concept of space as such, the ideal corporeal forms—point, straight line, plane, segment—being sufficient for its needs. On the other hand, space as a whole, as conceived by Descartes, was absolutely

> necessary to Newtonian physics. For dynamics cannot manage with the concepts of the mass point and the (temporally variable) distance between mass points alone. In Newton's equations of motion, the concept of acceleration plays a fundamental part, which cannot be defined by the temporally variable intervals between points alone. Newton's acceleration is only conceivable or definable in relation to space as a whole. Thus to the geometrical reality of space a new inertia-determining function was added.[61]

Hence, in our terms, Cartesian symbolic space—or numerically coordinatized space—furnishes the continuum of points at rest requisite to the formulation of Newtonian physics. The reader familiar with the history of general relativity will no doubt be reminded of the trouble coordinates gave Einstein in the so-called hole argument, with different solutions to the gravitational field equation applying at the same point in the same coordinate system.[62] Einstein's resolution of the hole argument involved the reversal or "deconstruction," if you will, of physically reified Cartesian symbolic space:

> On the basis of the general theory of relativity ... space as opposed to "what fills space," which is dependent on the coordinates, has no separate existence.... If we imagine the gravitational field, i.e. the functions g_{ik}, to be removed, there does not remain a space of type (1) ["flat space" or "Minkowski space"], but absolutely *nothing*, and also no topological space. For the functions g_{ik} describe not only the field, but at the same time also the topological and metrical structural properties of the manifold. A space of type (1), judged from the standpoint of the general theory of relativity, is not a space without a field, but a special case of the g_{ik} field, for which—for the coordinate system used, which in itself has no objective significance—the functions g_{ik} have values that do not depend on the coordinates. There is no such thing as an empty space, *i.e.*, a space without a field. Space-time does not claim an existence of its own, but only as a structural property of the field.[63]

While the passage above, from the 1952 appendix to Einstein's popular book on relativity, is generally regarded as referring specifically to the hole argument, it is difficult to speculate on whether Einstein had the hole argument in mind also in his discussion of the concept of space in the 1930s essays. Nevertheless, I cannot resist a comparison with Klein writing around the same time (1932):

> Only in this way [identifying the object of a second intention with the object of a first intention] can we come to understand that Descartes' concept of *extensio* identifies the extendedness of extension with extension itself. Our present-day concept of space can be traced directly back to this. Present-day Mathematics and Physics designate as "Euclidean Space" the domain of symbolic exhibition by means of line-segments, a domain which is defined by means of a coordinate system, a relational system [*Bezugssystem*], as we say nowadays. "Euclidean Space" is by no means the domain of the figures and structures studied by Euclid and the rest of Greek mathematics. It is rather only the symbolic illustration of the *general character of the extendedness* of those structures. Once this symbolic domain is identified with corporeal extension itself, it enters into Newtonian physics as "absolute space." At the present time it is being criticized by relativity theory, which has been steered by the question of "Invariance" into *trying to break through these symbolic bounds, while continuing to use this very symbolism* [my italics].[64]

It is doubtful, at least as regards the concept of Minkowski spacetime, that we can break through these symbolic bounds while continuing to use the very same symbolism.

Having achieved a grasp of the conceptual structure of modern symbolic algebra sufficient for our purposes, we are ready to proceed in the next chapter to the assimilation of that algebra into mathematical physics. With respect to the subsequent development of mathematical physics, we shall see that just as Vietan-Cartesian algebraic letter signs are, through the rules of the symbolic calculus, merged in meaning with the numerical quantities they designate, so in the equations or "formulas" of modern mathematical physics algebraic symbols are gradually merged in meaning with the physical quantities they represent.[65] The merging of symbol and quantity comes through most clearly in the multiplication and division of heterogeneous physical quantities, where the arithmetical operations can only have a symbolic intelligibility. The algebraic formula $p = mv$, for instance, takes mv as a physical quantity (momentum), but the operation of multiplication can be carried out solely on symbolic quantities in the Vietan-Cartesian sense. For what could it mean to take three kilograms, for instance, four meters per second times?

Historical accounts of the assimilation of algebra into modern mathematical physics as a rule presuppose the modern symbolic concept of number, and so we have at some length considered modern symbolic algebra from the perspective of some key stages in its historical development. This was necessary preparation for the clarification of the algebraic structure of the concepts of modern mathematical physics in the next chapter: for it is the Vietan-Cartesian algebra specifically that is assimilated into physics in the latter part of the seventeenth century through the eighteenth century. Having outlined the conceptual structure of this symbolic algebra we may proceed to the process of assimilation itself.

Notes

1. Newton 1769 [1707], 470.
2. Newton 1769 [1707], 11.
3. That modern mathematical physics is essentially determined by its symbolic mode of cognition is a thesis put forth by Jacob Klein in his classic *Greek Mathematical Thought and the Origin of Algebra*: "The intimate connection of the formal mathematical language with the content of mathematical physics stems from the special kind of conceptualization which is a concomitant of modern science and which was of fundamental importance in its formation" Klein 1992 [1934–1936], 4. Klein elaborates this thesis in a number of essays, particularly "The World of Physics and the 'Natural' World" (Klein 1985, 1–34).
4. Klein 1992 [1934–1936]. The definitive study of Klein is Hopkins 2011.
5. Euclid 1956, 2:277.
6. See Klein 1992 [1934–1936], Chapter 6, on the Greek understanding of number. With respect to the pure units of scientific arithmetic, Klein argues that the Platonic notion of separately existing pure units impeded the development of theoretical logistic or the science of arithmetical calculation, since the indivisibility of the pure units precluded calculation with fractions. According to Klein, Aristotle's critique of the Platonic separation thesis thus opens the possibility for theoretical logistic, in that Aristotle's pure units are abstracted measures that can be changed at liberty. For our purposes nothing hinges on the particular ontological conception of the pure units. Klein stresses that the concept of the pure unit itself is metaphysically neutral in Greek thought, in the sense that it "precede[s] all the possible differences of opinion regarding the *mode of being* of the 'pure' number units themselves ..." (54).
7. Euclid 1956, 2:278.
8. Klein 1992 [1934–1936], 4–5 and 147–149.
9. Diophantus 2009, 129.
10. Heath 2009, 52.
11. Bashmakova 1977 [1972], 6.
12. Diophantus 2009, 131.
13. Diophantus 1893, 1:6.

14. Diophantus 1893, 1:276.
15. Klein 1992 [1934–1936], 133–135.
16. Klein 1992 [1934–1936], 144–145.
17. Vieta 1992 [1646], 319.
18. Diophantus 1893, 9.
19. Vieta 1992 [1646], 328.
20. Vieta 1992 [1646], 345.
21. Vieta 1992 [1646], 321.
22. Vieta's emphasis on the "tediousness" of the ancient procedure suggests, however, that he is interpreting the *arithmos* of Diophantus in terms of his own symbolic conception of number such that the difference is a simply a matter of degree of generality rather than change in the conception of number itself.
23. Vieta 1992 [1646], 324.
24. Euclid 1956, 114. Vieta quotes the Aristotelian philosopher Adrastus' (second-century A.D.) commentary on Euclid's definition to the effect that "it is impossible to know how heterogeneous magnitudes may be conjoined" (Vieta 1992 [1646], 325).
25. Vieta 1992 [1646], 324.
26. Third precept of Vieta's rules for reckoning by species (*Introduction*, Chapter 4): "The denominations of products made by magnitudes ascending proportionally from genus to genus are related to one another in precisely the following way: A side multiplied by itself produces a square," and so one through the various powers (Vieta 1992 [1646], 334). Therefore magnitudes of heterogeneous dimension do have a ratio to one another.
27. Vieta 2006 [1646], 8.
28. Kline 1972, 279.
29. "When the equation of the magnitude which is being sought has been set in order, the rhetic or exegetic art, which is to be considered as the remaining part of the analytical art and as one which pertains principally to the application of the art (since the two others [zetetics and 'poristic'] are concerned more with general patterns than with precepts ...), performs its function both in regard to numbers if the problem concerns a magnitude that is to be expressed as a number, and in regard to lengths, surfaces, and solids if it is necessary to show the magnitude itself" (Vieta 1992 [1646], 346).
30. "And so, a proportion can be called the composition of an equation, an equation the resolution of a proportion" Vieta 1992 [1646], 324. That is, by cross multiplication, if $y : x^2 :: x^2 a$, for instance, then $ay = x^4$.
31. Kline 1972, 264–265.
32. Vieta 1992 [1646], 349.
33. Klein 1992 [1934–1936], 173–174.
34. Klein 1992 [1934–1936], 174.

35. Stevin 1585. Although Stevin's algebra has chronological precedence over Vieta's, Klein gives Vieta priority as the inventor of the first completely general symbolic calculus.
36. Stevin 1585, 495.
37. Stevin 1585, 501–503.
38. Stevin 1585, 495.
39. Klein 1992 [1934–1936], 191–192. Stevin's line of argumentation is reminiscent of the amusing observation, by the eponymous Parmenides in Plato's dialogue, that "mastery itself" must contain only mastery and so is related solely to "slavery itself," not to "a man" (who presumably possesses other attributes than being a slave). Similarly, since "number" for Stevin contains solely the numerical, any part of number is also number. The difference is that Plato's Forms are beings, while Stevin's "number" is a reified concept.
40. Vieta 1992 [1646], 324–325.
41. Descartes 1985–1991 [1637], 1:120–121.
42. Descartes 2001 [1637], 177.
43. Grosholz 2007, 167.
44. Descartes 1954 [1637], 19–20. I have translated this passage myself, as both the Smith and Latham translation (Descartes 1954) and the Olscamp translation (Descartes 2001) improbably render it as what "caused" the ancients to use arithmetical terms in geometry.
45. The "locus problem" is to find a locus of points from any one of which lines can be drawn intersecting given lines at a given angle, such that figures constructed from those lines have a fixed ratio.
46. There was precedent of a sort for nominalizing of ratios as numbers in the medieval practice of regarding the numerical quotient of the numbers comprising a ratio as the "denomination" of the ratio (Roche 1998, 46–47).
47. Klein 1992 [1934–1936], 220.
48. Weyl 1922, 8.
49. Descartes 1985–1991 [1637]: 1:121.
50. Descartes 1985–1991 [ca. 1628]:1:58.
51. Descartes 2001 [1637], 178–179.
52. Descartes 2001 [1637], 178.
53. The medieval algebraist Omar Khayyam too used multiplication by the unit to achieve homogeneity of dimension, but specifically for geometrical figures used to obtain algebraic solutions. Unlike Vieta and Descartes, however, Khayyam imposed no homogeneity requirement on algebraic equations themselves. See Jeffrey Oaks, "Al-Khayyām's Scientific Revision of Algebra." 2011. Accessed July 11, 2017. https://www.researchgate.net/publication/265780088_Al-Khayyam's_Scientific_Revision_of_Algebra.

54. Klein 1992 [1934–1936], 206*ff.* See also "The World of Physics and the 'Natural World,'" Klein 1985, 19–21. In the latter essay Klein attributes this assumption to Fermat as well.
55. Descartes to Mersenne, 27 July 1638. Descartes 1985–1991, III:119.
56. Klein 1992 [1934–1936], 211.
57. Einstein, "The Problem of Space, Ether, and the Field in Physics," Einstein 1982, 276–285.
58. Einstein, "Physics and Reality," Einstein 1982, 290–323.
59. Einstein 1982, 297.
60. Einstein 1982, 279.
61. Einstein 1982, 279–280.
62. There exists a huge body of literature on the hole argument. Two good basic accounts are Norton 1987 and Stachel 1989b.
63. Einstein 1961 [1916], 176.
64. Jacob Klein, "The World of Physics and the 'Natural World'," in Klein 1985, 21.
65. Newton is a noteworthy exception, for he actively resisted numerical methods and employed a non-numerical geometrical algebra in physics. However, historically the main current of algebraic development in modern physics is Vietan-Cartesian and numerical. Moreover, the fact that Newton resisted numerical methods in physics does not preclude his having operated, as Klein maintains, with a Cartesian concept of space.

CHAPTER 5

The Historical Sense-Structure of Modern Algebraic Physics

The assimilation of Cartesian algebra into mathematical physics met with significant opposition. We have already noted Newton's objections to algebra in geometry, and as late as 1798, Laplace feels the need to justify the use of algebra in theoretical physics. Real issues of physical intelligibility engendered the resistance to the use of algebra in physics, not merely mathematical conservatism or an exaggerated reverence for the ancients. We touched on some of these issues last chapter. In this chapter we shall consider more closely those historical developments by which Cartesian algebra came to be the principal, and by now the exclusive, mathematical and conceptual language of theoretical physics. Fortunately, a body historical scholarship in this area already exists—if not always from the perspective of our present concerns—and so I shall avail myself of it freely in what follows.[1] The present chapter offers not a historical narrative or even a historical sketch, but rather a discussion of some emblematic examples, for the purpose of identifying the decisive conceptual moves underlying the algebraic structure of concepts in modern physics.

5.1 Pre-Algebraic Physics in Galileo

While Galileo is rightly regarded as the founding father of modern mechanics, his mathematical method is essentially Euclidean (with a noteworthy exception to be discussed below). Galileo neither uses equations nor defines physical concepts or principles numerically, although he necessarily

J. K. Cosgrove, *Relativity without Spacetime*,
https://doi.org/10.1007/978-3-319-72631-1_5

interprets his experimental results in terms of numerical measurements. In this respect, Galileo's science comports with the seventeenth-century understanding of rigorous science in general, which, as such, must respect the fundamental Greek distinction between continuous and discrete quantity. Euclidean rigor in Galileo's case is accomplished by means of representing physical quantities geometrically as line lengths, such that the Euclidean method of ratio and proportion may be applied indirectly to physical quantities of interest.

The one exception to Euclidean rigor is Galileo's method of compounding ratios of heterogeneous quantities, a departure from the Euclidean law of homogeneity for ratios. Recall that compounding ratio $B : A$ with ratio $C : B$ yields $C : A$ for the compound ratio. In view of Euclid's definition, according to which a ratio is a relation between magnitudes of the *same kind*, all the quantities involved in the ratios to be compounded must be of the same kind, for otherwise a heterogeneous ratio would result. Euclid never compounds a ratio of lines, for instance, with a ratio of planes; for the compounded ratio then would be the ratio of a line to a plane, and a plane cannot be a multiple of the length of a line. By contrast, a proportion between ratios of heterogeneous quantities is perfectly acceptable—for example, a ratio of lines set in proportion to a ratio of planes—since such a proportion involves no ratio between unlike quantities.

Galileo gets around the homogeneity requirement for ratios by representing all the quantities as line lengths and then compounding ratios of line lengths. Proposition IV, Theorem IV of Day Three in Galileo's *Two New Sciences*, for instance ("On Equable Motion"), reads: "If two moveables are carried in equable motion but at unequal speeds, the spaces run through by them in unequal times have the ratio compounded from the ratio of the speeds and from the ratio of the times."[2] Galileo designates line lengths A and B respectively as the speeds and line lengths C and D as the times, yielding $A : B$ for the speeds and $C : D$ for the times. If we appropriately adjust the line lengths such that B equals C, as we may, then we obtain $A : D$ for the compound ratio of interest, itself proportional to the ratio $G : L$ of the distances traversed. But line length A represents speed and line length D time, and so the law of homogeneity for ratios has been violated indirectly. While there is precedent in medieval science for compounding heterogeneous ratios, the practice nevertheless betrays a certain lack of rigor.[3] We can restore homogeneity, however, by interpreting Galileo's proposition in terms of two proportions or a "conjoint proportion" rather than a single proportion. Instead of a single proportion, that is, between the compounded ratio of the speeds and the times and the ratio of the distances—$G : L :: (A : B)\,(B : D)$—we instead regard the ratio of the distances as proportional conjointly to the ratio of the speeds and to the ratio of the times: $G : L :: A : B$ (with $C : D$ held constant) and $G : L :: C : D$ (with $A : B$ held constant). The technically illicit compounding may thus be regarded as an abbreviation for two proportions.

Another instructive example for our purposes is Galileo's law of free fall (Day Three, Proposition II, Theorem II), which we are used to expressing algebraically as $s = \frac{at^2}{2}$. Here is how Galileo presents it: "If a moveable descends from rest in uniformly accelerated motion, the spaces run through in any times whatever are to each other as the duplicate ratio of the times; that is, the squares of those times."[4] To demonstrate the proposition we start with the medieval mean speed theorem, which Galileo has proved in the previous proposition: the distance traversed by a uniformly accelerated body is equal to the distance traversed in an equal time by a body moving uniformly at one-half the terminal speed of the accelerated body. And, since by definition a uniformly accelerated body acquires equal increments of velocity in equal times, the ratio of distances traversed is proportional twice over to the ratio of times— that is, it is proportional conjointly to the ratio of terminal velocities, which itself is proportional to the ratio of times, and to the ratio of times again. In this case, though, since our quantities are homogeneous, we can express ourselves more compactly by compounding the ratio of times with itself ("duplicating the ratio").

Galileo consequently gives the law of fall as, "The ratio of distances is as the duplicate ratio of times," adding, in accordance with accepted usage: "... that is, as the squares of those times." Here we have no multiplication of t by itself to yield t^2, but rather a duplicate ratio of times, which, given the representation of those times by line lengths, and given also that the size of a square increases as the duplicate ratio of its side, may be called the "squares of those times." Hence, Galileo's proportion is: $s_1 : s_2 :: (t_1 : t_2)\,(t_1 : t_2)$ (with s representing distance and the parentheses designating the operation of compounding rather than multiplication).

We thus encounter in Galileo a non-numerical conception of physics in strict accord with the requirements of its subject matter: continuous physical quantity. Moreover, we have seen that, with the exception of the compounding of heterogeneous ratios, Galileo employs a direct method of representation. That is to say, Galileo's mathematical operations are directly intelligible in terms of ratios and proportions and require no translation into physically intelligible language. His symbolic line lengths are merely a visual aid and do not enter into the physical meaning of his theorems in physics: in principle they are dispensable. That is why Galileo gives his theorems, the law of fall for instance, in a direct language free from mathematical symbolism ("The ratio of distances is as the duplicate ratio of times").

Galileo's non-symbolic conception of mathematical representation in physics raises the question of how the symbolic-algebraic method of Descartes can be integrated into Galilean physics. For only the assimilation of Cartesian algebra into physics yields "modern physics" in the sense we understand it today. Yet the displacement of the Euclidean mathematics of ratio and proportion by Cartesian algebra renders the proximate object of modern theoretical physics a symbolic construction. Thus, according to Hermann Weyl, for instance, our intuitions of space and time must "give way to *symbolic construction* of exactly the same kind as that which Hilbert carries through in mathematics."[5] The idea of mathematical science as symbolic construction is commonplace enough in contemporary history and philosophy of science, but it has not as a rule been understood in the context of the prior symbolic construction of number as we focused on last chapter; rather, in such analyses the modern symbolic concept of number is simply taken for granted. But it is the Vietan-Cartesian concept of number that renders possible the very symbolic construction of natural science to which Weyl alludes in the quotation above. As a result, the relationship between symbolic construction and the physical world is left unclarified; just as is, for instance, the relationship between "the number 4" itself and these four apples in front of me.

5.2 The Assimilation of Algebra into Physics

The process by which symbolic algebra was assimilated into physics spanned the latter part of the seventeenth and most of the eighteenth centuries.[6] Near the end of this development, the illustrious mathematician Pierre Simon de Laplace, in a remarkable passage from the introduction to his *Celestial Mechanics* of 1798, lays out with exceptional explicitness, if not equal conceptual clarity, the symbolic-algebraic structure of concepts thus bequeathed to subsequent mathematical physics:

> In uniform motions, the spaces described are proportional to the times. But the times employed in describing a given space are longer or shorter according to the magnitude of the moving force. From these differences has

> arisen the idea of *velocity*, which, in uniform motions is the ratio of the space to the time employed in describing it. Thus s representing the space, t the time, and v the velocity we have $v = \frac{s}{t}$. Time and space being heterogeneous and consequently not comparable quantities, a determinate interval of time, such as a second, is taken for a unit of time and in like manner a portion of space, such as the meter for the unit of space, and then time and space become abstract numbers, which express how often they contain units of their species [how many units they contain], and thus they may be compared with one another. By this means the velocity becomes the ratio of two abstract numbers, and its unit is the velocity of a body which describes a meter in one second.[7]

The fact that Laplace feels compelled to justify the formula $v = \frac{s}{t}$, a century after it was first introduced, is noteworthy in itself. He initially characterizes uniform velocity in a Galilean manner, in terms of Euclidean proportion. Note in particular that Laplace does *not* say that in uniform motion "distance is proportional to time"—our modern-day algebraic conception of a proportion with single quantities related by a constant of proportionality—but rather that "the spaces described are proportional to the times": a proportion between homogeneous *ratios*. But almost immediately he apparently reverses course with the novel assertion that velocity is in fact the ratio of the space to the time, expressing that ratio algebraically as the fractional number $\frac{s}{t}$. Then, in a last echo of the classical understanding of ratio, Laplace recalls that space and time cannot really be placed in ratio, nor can a quantity of space be divided by a quantity of time: for time and space are "heterogeneous and consequently not comparable quantities." And so instead of a ratio between space and time we designate units of space and time respectively, such that space and time become "abstract numbers" upon which we can now carry out the operation of division.

There is much more to comment upon in the quoted passage, but of utmost interest for us is how the introduction of unit quantities of space and time renders space and time abstract numbers. Clearly, by "abstract number" Laplace understands number in the Vietan-Cartesian symbolic sense, not "numbers of" pure units in the manner of Euclid or Diophantus. But if we understand s and t as symbolic dimensionless numbers, such that s can be divided by t, what makes $v = \frac{s}{t}$ an equation of physics, relating physical quantities of space and time to the physical quantity of velocity?[8] It is the stipulation of units that allows the letter symbols to represent simultaneously purely symbolic numbers and "numbers of" the designated physical measures. Thus the concept of velocity in Laplace's formula $v = \frac{s}{t}$ precisely parallels in structure the concept of number in Vieta and Descartes. Just as Vieta's species are at once generalized magnitudes and "numbers of" homogeneous units (the "homogeneous element" of the equation), so too do Laplace's algebraic variables at once represent generalized magnitudes ("abstract numbers") at the same time they function as "numbers of" the physical unit measures. Moreover, this dual mode of intentionality—what Klein terms the identification of the object of a second intention with the object of a first intention—is accomplished in the same manner in both cases: the letter symbol is merged in meaning with the quantity it represents by means of the rules which govern the symbolic calculus. Only in this manner can Laplace's s and t be both "abstract numbers" and numbers of physical units.[9]

A generation earlier (1765), d'Alembert had with somewhat more clarity stressed that the physical meaning of $v = \frac{s}{t}$ is "[nothing] other than that the velocities of two bodies are to each other as the quotients of the spaces divided by the times, provided one represents the spaces and times as abstract numbers which bear to one another the same ratio as these spaces and these times."[10] John Roche, in his account of the assimilation of algebra into mathematical physics, notes that in eighteenth-century physics the equal sign was often interpreted as an abbreviation for a proportion or "ratio equation" and did not necessarily designate an equality of quantities.[11] D'Alembert again writes: "Sometimes, in geometry and in mechanics, that is called an equation which is nothing other than a simple proportionality indicated in an abbreviated manner."[12]

How would we render Laplace's $v = \frac{s}{t}$ intelligible in terms of Galileo's theorem on equable motion discussed above in Sect. 5.1? In most rigorous form, without the employment of heterogeneous compounded ratios, Galileo's theorem says that the ratio of distances is jointly proportional to the ratio of the speeds and to the ratio of the times. Less rigorously we can compound the ratios of speeds and times as Galileo does, with $s_1 : s_2 :: (v_1 : v_2)\,(t_1 : t_2)$. Now let the variables s, v, and t represent continuous numerical values of the measures of distance, speed, and time respectively. Designating the consequents of these ratios as units and multiplying through the compounded ratios we obtain $s : 1 :: (vt : 1)$. We can then either directly nominalize the ratios as fractional numbers or apply the rule of cross-multiplication to resolve the proportion into the equation $s = vt$ and, therefore, $v = \frac{s}{t}$. Following d'Alembert's lead, instead of regarding $\frac{s}{t}$ as the "division" of a quantity of space by a quantity of time, a physically unintelligible operation in any case, we instead interpret the equation $v = \frac{s}{t}$ as a symbolic translation of the original Galilean proportion. Physical sense thus accrues indirectly to the symbolic construction $\frac{s}{t}$. There is no actual division of s by t, but rather a direct proportion of velocity to distance and an inverse proportion of velocity to time. The division sign represents the two proportions economically in a single algebraic expression.

Even more revealing in some respects is the algebraic translation of Galileo's law of free fall into the formula $s = \frac{at^2}{2}$. For two fixed accelerations, the ratio of distances is proportional jointly to the duplicate ratio of the times and the ratio of accelerations, since the ratio of accelerations is proportional to the ratio of terminal velocities. We can compound these ratios to obtain $s_1 : s_2 :: (t_1 : t_2)\,(t_1 : t_2)\,(a_1 : a_2)$. However, since one of the ratios of times enters the proportion by means of the mean speed theorem, according to which a uniformly accelerated body traverses a distance equal to half the distance it would cover at terminal speed, we insert the additional ratio 1:2 in our compounded ratio, yielding $s_1 : s_2 :: (t_1 : t_2)\,(t_1 : t_2)\,(a_1 : a_2)\,(1 : 2)$. Now, in our usual manner, we designate the consequents s_2, t_2, and a_2 respectively as units of distance, time, and acceleration and then multiply through to obtain $s : 1 :: (t)\,(t)\,(a) : 2$. The proportion accordingly resolves to the equation $s = \frac{at^2}{2}$, in accordance with Vieta's dictum that a proportion composes an equation and an equation resolves a proportion.

In comparing proportions with algebraic equations we must bear in mind this decisive difference: proportions equate *relations* (ratios) between quantities, while equations equate quantities themselves. Therefore, as we have seen, when a proportion is resolved into an equation the original ratios are nominalized as if they were quantities in their own right—artifactual quantities, if you will. Clearly, t^2 in the example above represents no real physical quantity but rather a nominalized compound ratio. Let us designate such pseudo-quantities, which are generated whenever proportions involving ratios of physical quantities are resolved into algebraic equations, as "symbolic quantities." Hence, the *mv* in the momentum formula $p = mv$ is a symbolic quantity in the same sense as t^2 in Galileo's the law of fall. From this perspective the nineteenth century American philosopher J.B. Stallo rightly objects, in his *Concepts and Theories of Modern Physics* (1881), to the

> error respecting the true nature of arithmetical and algebraic quantities [which] has become next to ineradicable by reason of the inveterate use of the word "quantity" for purposes of designating indiscriminately both extended objects or forms of extension and the abstract numerical units or aggregates by which their metrical relations are determined … The use of letters as algebraic symbols, i.e., as representatives of numbers, is in itself a serious (though, perhaps unavoidable) infirmity of mathematical notation. In the simple formula, for instance, expressive of the velocity of a moving body in terms of space and time ($v = \frac{s}{t}$), the letters have a tendency to suggest to the mathematician that he has before him direct representatives of the things or elements with which he deals, and not merely their ratios expressible in terms of numbers. In every algebraic operation the use of letters obscures the real nature, both of the processes and the results, and tends to strengthen ontological prepossessions.[13]

Both "processes and results" are obscured precisely because we fail to attend to the historically constituted meaning-structure of algebraic equations in theoretical physics.

5.3 A Case Study: Newton and "Quantity of Motion"

We conclude our discussion of the historical assimilation of algebra into physics with a brief case study: Newton's treatment of "quantity of motion" (momentum) in the *Principia*. The case is especially informative in that, generally speaking, Newton is far more sensitive to issues of algebraic representation in physics than are his immediate predecessors. The sensitivity evidently increased as Newton's career progressed, and it comes out particularly in some of the revisions to the *Principia* in the third edition of 1726. Before going into these revisions, we recall that already in his *Universal Arithmetic* (published anonymously in 1707 but based on work from the 1670s) Newton expressed his increasing dissatisfaction with algebraic methods in geometry, over against the more rigorous geometry of the ancients. Henry Pemberton, who supervised the preparation of the third edition of the *Principia*, reports in 1728 that

> I have often heard him censure the handling geometrical subjects by algebraic calculations; and his book of Algebra he called by the name of Universal Arithmetic, in opposition to the injudicious title of Geometry, which *Descartes* had given to the treatise, wherein he shows, how the geometer may assist his invention by such kind of computations.... Of their [the ancients] taste, and form of demonstration, Sir ISAAC always professed himself a great admirer: I have heard him even censure himself for not following them yet more closely than he did; and speak with regret of his mistake at the beginning of his mathematical studies, in applying himself to the work of *Descartes* and other algebraic writers, before he had considered the elements of *Euclid* with that attention, which so excellent a writer deserves.[14]

It is of no minor significance that Newton's remarks above to Pemberton are from around the time of the preparation of the third edition of the *Principia* (1726), for they give us a measure of insight into the motivation for some of the changes he made to that text.

At the beginning of *Arithmetica* Newton distinguishes between "computation by means of numbers, as in vulgar arithmetic, or by species, as usual among algebraists."[15] Then in the immediately following section on the definition of terms, he remarks that "by number we understand, not so much a multitude of unities, as the abstracted ratio of any quantity, to another quantity of the same kind, which we take for unity."[16] Newton's definition of number is essentially the one already proposed by Wallis (4.4.2 above), and among such numbers, in the sense of abstracted ratios, are integers, fractions, and surds (irrationals). Clearly, the notion of "abstracted ratio" is a development of Vieta's symbolic number concept, but observe that for Newton the distinction between "vulgar" numbers and species falls wholly within the symbolic category. Unlike Vieta himself, for instance, Newton does not interpret the individual numbers of the ancients as symbolic entities assigned determinate values (e.g., "3" instead of "*x*"), but instead distinguishes the ancient concept of number ("multitude of unities") from the modern Vietan-Cartesian concept ("abstracted ratio"). The latter distinction applies to both determinate numbers ("vulgar arithmetic") and indeterminate numbers (algebra or computation by "species"). Newton's superior insight into the definition of number is thus evinced in his declining to attribute to the ancients the modern understanding of "vulgar numbers." He is fully aware of the meaning of *arithmoi* in the ancient Greek sense and does not reduce it to a "less abstract" version of our own concept of number.

If the methods of the ancients are truly more rigorous, however, then a rigorous method in physics cannot be numerical, for the ancient concept of number (multitude of units) cannot accommodate the continuous quantities of physics. Number as "abstracted ratio" allows for the direct conversion of ratios to fractions, such that proportions can be rewritten as algebraic equations. Nevertheless, on Newton's expressed preference for the ancients the calculational superiority of Cartesian algebra is not to be mistaken for an advance in mathematical rigor. Niccolò Guicciardini aptly characterizes Newton's aversion to algebraic methods in terms of an increasing concern over "referential content":

> The importance of adopting a mathematical method endowed with referential content was particularly relevant for Newton's science of motion. In the seventeenth-century, the idea that the language for natural philosophy had to be geometrical was deeply rooted. Since Galileo's time, the "Book of Nature" had been thought to be written in "circles and triangles and other geometrical figures"; it was not, however, written in algebraic symbols.... Newton often characterized the symbolic methods of algebra and calculus as merely heuristic tools devoid of scientific character.[17]

It is in light of this expressed dissatisfaction with algebraic methods around the time of the third edition of the *Principia* that we shall interpret Newton's specific alterations therein to the description of quantity of motion.

We noted in Chap. 2 (Sect. 2.4) that Newton was careful to distinguish between the qualitative definition of a physical quantity and its quantitative measure. He defines "impressed force," for instance, as an action on a body which changes its state of motion (Book I, Definition IV), while the quantitative measure of impressed force is given in the second law of motion as a proportion, in terms of its effect: "The change of motion is proportional to the motive force impressed."[18] Hence, what we express with the formula $F = ma$ is not intended by Newton as a definition of impressed force, even if the equal sign in our algebraic version may well suggest the word "is." Yet Newton does define quantity of motion quantitatively as a proportion: quantity of motion is "the measure of the same, arising from the velocity and quantity of matter conjointly."[19] Our own concept of momentum perhaps can be thought of qualitatively in terms of how much effort it takes bring something to a stop (thus the use of momentum as a metaphor in sports and politics), but Newton's quantity of motion is obviously *already* a quantification. Thus what would really require qualitative definition would be motion itself. Roche plausibly suggests that Newton thought of motion as one of those quantities, like absolute time, "to which we have no direct access and can only know only indirectly as measures."[20]

In his commentary on the definition Newton at least is clear that by the phrase "arising from the velocity and quantity of matter conjointly" he intends a proportion: "Therefore, in a body twice as large, with equal velocity, it is double, and with double velocity, quadruple" ("*ideoque in corpore duplo maiore, aequali cum velocitate, duplus est, et dupla cum velocitate quadruplus*").[21] Here Newton's commitment to Euclidean rigor, expressed in the opening quotation of this chapter, comes through with exceptional clarity in that Newton gives the definition in terms of a "conjoint proportion" ("*orta ... conjunctim*") rather than a heterogeneous compounded ratio ("*ratione composite*") of velocities and quantities of matter. That is, declining Galileo's method of heterogeneous compounded ratios, Newton specifies instead two proportions joined together: (1) a proportion between quantities of motion and quantities of matter with velocity held constant, and (2) a proportion between quantities of motion and quantities of velocity with quantity of matter held constant.[22]

Nevertheless, Newton is often willing to relax Euclidean rigor, especially when it comes to calculations—albeit preferably with appropriate disclaimers. Thus in the first two editions of the Principia (1687 and 1713), in the Scholium to the Laws of Motion, he gives the quantity of motion as the product of quantity of matter and velocity: "*Tandem ducendum erit corpus A in chordam arcus TA, quae velocitatem ejus exhibit.*"[23] While Newton's use of "*ducendum ... in*" in preference to the more unambiguously arithmetical "*multiplicare*" suggests the construction of a rectangle by "drawing" (*ducere*) body *A*'s mass ("quantity of matter") into its velocity, the same terminology can also carry a numerical sense, which is evidently the case here.[24] That is, Newton does not speak here of a proportion between ratios but rather a single [quantity of] motion—"its motion" ("*motus ejus*")—at a point. Thus we essentially have a numerical representation of quantity of motion as $Q = mv$. In the third edition of the *Principia*, however, around the time he was expressing to Pemberton his aversion to algebra, Newton inserts the qualifying phrase "if I may say so" or "so to speak" ("*ut ita dicam*"): "*Tandem ducendum erit corpus A (ut ita dicam) chordam arcus TA, quae velocitatem ejus exhibit.*"[25] Evidently Newton intends to underline that the "product" of mass and velocity is to be taken as an abbreviation for the proportion he had already given in Definition Two. Newton here confirms, therefore, what d'Alembert would later insist upon, that the physical meaning of algebraic representations in physics is "nothing other" than that which is given by proportions.

Another text of interest with regard to numerical methods in physics is Definition VIII on the "motive quantity of centripetal force." Newton writes in 1687 and 1713,

> Therefore the accelerative force is to the motive force as the velocity is to the motion. For the quantity of motion arises from the velocity drawn into ["*ducta in*" or "multiplied by"] the quantity of matter, and the motive force [arises from] the acceleration drawn into the same quantity of matter. (*Est igitur acceleratrix ad vim motricem ut celeritas as motum. Oritur enim quantitas motus ex celeritate ducta in quantitatem materiae, & vis motrix ex acceleratrice ducta in ejusdem materiae*).[26]

But in 1726 Newton replaces "ducere in" with the more rigorous "conjoint proportion":

> For the quantity of motion arises from the velocity and quantity of matter conjointly, and the motive force [arises from] the acceleration and the same quantity [of matter] conjointly. (*Oritur enim quantitas motus ex celeritate & ex quantitate materiae, & vis motrix ex acceleratrice & ex quantitate ejusdem materiae conjunctim*).[27]

Thus Newton's remarks to Pemberton on algebra around the time of the preparation of the third edition are once again reflected in an actual revision to the text.

These changes in the third edition are revealing of Newton's progressive aversion to algebra in physics. There can be no doubt that Newton regarded algebraic representation as an abbreviation for ratios and proportions and did not invest such representations with direct physical intelligibility. This is in keeping with the general view of the matter in seventeenth and eighteenth century physics.[28] A notable exception, which we shall see is highly relevant to the physical interpretation of Minkowski spacetime, is conservation laws. The law of conservation of quantity of motion for elastic collision, for instance, can be written algebraically as

$m_1 v_1 + m_2 v_2 = m_1 v_{1_0} + m_2 v_{2_0}$. However, it is not the resolution of a proportion in Vieta's sense, since conservation laws in general equate sums of quantities rather than ratios of quantities. The interpretation of such an equation as the preceding in fact requires some care. In the present case, while the individual terms are indeed algebraic abbreviations for compounded ratios, the equation as a whole is not an abbreviated proportion in the sense we have outlined above. Rather, to translate conservation laws out of algebra we have to express them verbally, as both Newton and Descartes before him do:

> The quantity of motion, which is determined by adding the motions made in one direction and subtracting the motions made in the other direction, is not changed by the action of bodies on one another. (Newton, *Principia*, Law III, Corollary III[29])

And likewise,

> God ... always preserves the same quantity of motion in the universe. (Descartes, *Principles of Philosophy*, Part Two, 36[30])

A conservation law, then, in its physical intelligibility, is neither an algebraic equality per se nor a proportion, but rather, like the Pythagorean Theorem itself, an equality of summed quantities. We shall have to determine into which category the Minkowski spacetime interval falls.

Notes

1. In this chapter I have benefited especially from John Roche's *The Mathematics of Measurement* (Roche 1998).
2. Galileo 1989 [1638], 151.
3. Roche 1998, 44.
4. Galileo 1989 [1638], 166.
5. Weyl 1949, 113.
6. See Roche 1998, chaps. 6 and 7.
7. Quoted in Roche 1998, 138.
8. The modern symbolic formalization of the operation of division is evident in the fact that nowadays we do not distinguish $\frac{a}{b}$ in the sense of "*a* divided by *b*" (how many times *b* goes into *a*) from $\frac{a}{b}$ in the sense of "*a*

divided into *b* parts." 6 divided by 3 is 2, for instance, since 3 goes twice into 6. Also, 6 divided into 3 parts makes 2, since 3 parts make half of 6 parts. But the distinction between the two cases does not register in the symbolically formalized operation $\frac{6}{3}$.

9. On this basis can we understand the propensity in the theory of relativity to regard the velocity of light *c* as a "dimensionless number." The original motivation for this procedure clearly is convenience in calculation, for if we at the same time designate *c* as unit velocity then *c* disappears from our equation and we do not have to write it down or calculate with it. But these two procedural steps (dropping the dimensions of *c* and adjusting it to unity) have no relation to one another. We would as well regard *c* as the dimensionless number 186,000, after all. But that would not serve the purpose of symbolic convenience, so we rid ourselves of the symbol *c* for notational purposes by regarding it as the number 1. The symbolic elimination of *c* is then regarded as an actual elimination, as if the physical dimensions were not still intended by the given symbolic expression. As we learned in Chap. 2, this can lead to great confusion. We imagine, for instance, that energy *is* mass ($E = m$) when all the equation $E = mc^2$ really entails is that energy is proportional to mass (and that compared with amounts of energy we are used to, the increment of energy associated with an increase in mass is very large).
10. Quoted in Roche 1998, 104.
11. Roche 1998, 126.
12. Quoted in Roche 1998, 126.
13. Stallo 1960 [1881], 275, 277.
14. Pemberton 1728. According to Guicciardini, Newton's preference for the geometrical methods of the ancients over modern algebraic methods "[has its] roots in the 1670s, becoming stronger as the years passed" (Guicciardini 2002, 318).
15. Newton 1769 [1707], 1.
16. Newton 1769 [1707], 2.
17. Guicciardini, 2004, 144.
18. Newton 1999 [1726], 416.
19. Newton 1999 [1726], 404.
20. Roche 1998, 107–108.
21. Newton 1972 [1726], 1:40.
22. The "holding constant" of the other ratio ("*aequali cum velocitate*") makes no sense for compound ratios, but it is necessary for a conjoint proportion since the two proportions cannot be given independently.The ratio of the quantities of motion, that is, is proportional to the ratio of the quantities of matter not *simpliciter*, but only on the condition that the ratio of velocities is held constant (and vice versa).

23. Newton 1972 [1726], 1:67.
24. Motte in 1729, based on the third edition, translates "*ducendum* ... *in*" as "product" (Newton 1846 [1726], 91) as does Cajori in his 1934 revision of Motte's translation (Newton 1934 [1726], 23). Motte in general seems to prefer the more literal "drawn into," but in this case it does not fit in with the grammar of the sentence. Cohen instead chooses "multiply" (Newton 1999 [1726], 425).
25. Newton 1972 [1726], 1:67.
26. Newton 1972 [1726], 1:45. In this case Motte keeps "*ducta in*" from the first and second editions and renders it more literally as "drawn into" (Newton 1846 [1729], 76), while Cajori has "multiplied by" (Newton 1934 [1726], 216). Cohen uses simply "arises from" in accord with Newton's removal of "*ducta in*" in the third edition (Newton 1999 [1726], 407).
27. Another instance of Newton's qualifying the phrase "*ducere* ... *in*" is Proposition LXXXVIII, Theorem XLV of Book One, where Newton adds in the third edition, "*si ita loquar*" (Newton 1972 [1726], 1:327).
28. According to Roche (1998, 108), "[t]hroughout the seventeenth and eighteenth centuries the product or division of physical quantities was often understood as an abbreviated statement of a compound ratio."
29. Newton 1999 [1726], 420.
30. Descartes 1985–1991 [1644], 1:240.

CHAPTER 6

Desedimentation of Minkowski Spacetime

We must now apply the historical considerations of the last two chapters to the concept of Minkowski spacetime. The Minkowski spacetime interval is sometimes called a "generalization" of the Pythagorean Theorem, but Einstein himself always more correctly referred to the *formal* analogy between the four-dimensional spacetime continuum and the three-dimensional continuum of Euclidean space:

> The four-dimensional mode of consideration of the "world" is natural on the theory of relativity, since according to this theory time is robbed of its independence.... But the discovery of Minkowski, which was of importance for the formal development of the theory of relativity, does not lie here. It is to be found rather in the fact of his recognition that the four-dimensional space-time continuum of the theory of relativity, in its most essential formal properties, shows a pronounced relationship to the three-dimensional continuum of Euclidean geometrical space.[1]

The theory of Minkowski spacetime stands or falls with the analogy between the interval and the Pythagorean Theorem. If we are to regard the spacetime interval as anything more than one member of the general Lorentz transformation $c^2t'^2 - x'^2 = c^2t^2 - x^2$, however, we must establish a more than merely formal analogy between the interval and the Pythagorean Theorem.

J. K. Cosgrove, *Relativity without Spacetime*,
https://doi.org/10.1007/978-3-319-72631-1_6

Clearly the relevant consideration for evaluating the Pythagorean analogy is the meaning of the squared terms. When we represent the Pythagorean Theorem algebraically, the squared terms ultimately represent actual geometrical quantities—the sizes of geometrical squares built on the sides of a right triangle. If the analogy with the Minkowski interval is to be regarded as more than merely formal, the squared terms in the expression $c^2dt^2 - dx^2$ must represent geometrical quantities as well. But how do we determine whether they do? Our preliminary analysis identified two levels of historically sedimented meaning in the equations of physics: (1) the equations of physics generally function as symbolic abbreviations for proportions, and (2) algebraic terms in the equations of physics often represent symbolic quantities (or pseudo-quantities) instead of real quantities. We recall, for instance, that t^2 in the algebraic version of Galileo's law of fall is no real quantity in nature but rather the symbolic abbreviation for a duplicate ratio, meaning that the length of free fall under gravity increases four times when the time of fall doubles.

We thus have a method by which to proceed: Should the expression $c^2dt^2 - dx^2$ reveal itself as the algebraic translation of a proportion, then the squared terms must be regarded as pseudo-quantities and any analogy to the Pythagorean theory such as could underwrite the concept of spacetime is fatally undermined. However, should the interval, like the Pythagorean Theorem itself or conservation laws and so forth, not find its origin in a proportion, that result would redound to the advantage Minkowski spacetime. For in that event, the spatial term at least would represent a geometrical square as in the Pythagorean Theorem. The time term c^2dt^2 would remain anomalous, though, for a square built on the distance light "would travel" in the time interval between two events clearly has no physical significance.

To determine the category to which the so-called spacetime interval belongs, we must translate the expression $c^2t^2 - x^2$ *out of algebra*. It will serve our purpose to dispense with spatial components, expressing the distance between events with the single variable *x*. For although the resolution of the spatial term into components renders the Pythagorean analogy more suggestive in visual-notational terms, the question of present interest is whether space and time can be merged into a single metrical continuum—and we already know the three dimensions of space are so united. Since $c^2t^2 - x^2$ can be rewritten as the product of factors $(ct + x)(ct - x)$, a proportion immediately suggests itself. Clearly, the Lorentz transformation in the form $(ct' + x')(ct' - x') = (ct + x)(ct - x)$, with some preliminary algebraic manipulation, may be composed as a proportion between compound ratios. However, we cannot in this manner truly discern the sense structure of the equation. Instead, in accord with our expressed aim of desedimenting the concept of Minkowski spacetime, we must proceed genetically or in the opposite direction. *That is, we start rather with a non-algebraic derivation of the* Lorentz *transformation and recover the original proportion from which it is derived, thereby reactivating the historically sedimented sense-structure of the algebraic resolution of the proportion.* In relativistic physics too, that is, to cite once again the prescient words of d'Alembert, "[T]hat is called an equation which is nothing other than a simple proportionality indicated in an abbreviated manner." The genetic procedure is all the more necessary in view of Minkowski's appropriation of an *already* historically sedimented development of meaning—Vietan-Cartesian symbolic algebra—as point of departure for the construction of the concept of four-dimensional spacetime. For our task is just to uncover the sense-genesis of the very algebraic expression that Minkowski presupposed as his starting point.

We set up a thought experiment as depicted in the spacetime diagram below (Fig. 6.1)[2]:

Two clocks A and B, with trajectories so designated (A coinciding with the time axis), intersect at the origin of coordinates. Sometime later, a

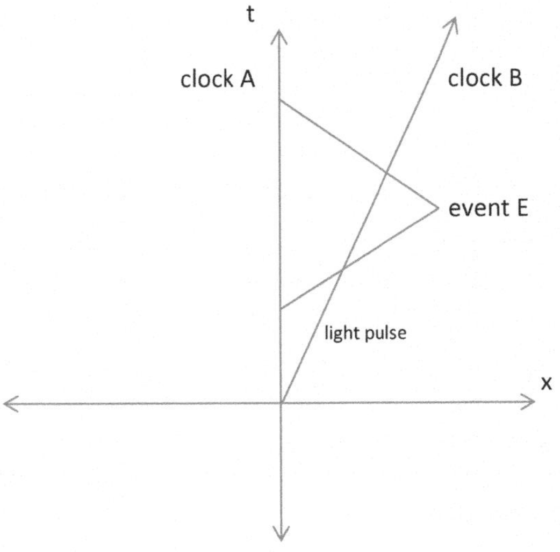

Fig. 6.1 Two world-lines with light pulse

light pulse departs from A and passes B on its way to reflecting off a mirror (event E) and subsequently arriving back at A. We designate the time interval between the clocks' initial rendezvous and event E as t_A and t_B appropriately (relative to inertial frames A and B respectively) and the distances as x_A and x_B respectively. We further designate the time it takes the light pulse to travel (in either direction) between A and E or between B and E as T_A and T_B respectively. Consequently, for the light pulse's trip to the mirror the time of transmission from A is $t_A - T_A$ and the time of reception at B is $t_B - T_B$. And by analogous considerations for the return trip the light pulse's time of departure from B is $t_B + T_B$ and the time of reception at A is $t_A + T_A$.

Considering first the light pulse's trip *to* the mirror, we form the ratio of the time of transmission at A and reception at B: $(t_A - T_A) : (t_B - T_B)$. Likewise, for the return trip we form the ratio of the time of transmission at B and reception at A: $(t_B + T_B)\ (t_A + T_A)$. And, since the situation is symmetrical as regards the "to" and "from" trips of the light pulse, the two ratios are proportional: $(t_A - T_A) : (t_B - T_B) :: (t_B + T_B)\ (t_A + T_A)$. We could call this the "pure form" of the Lorentz transformation, exclusively in terms of homogeneous ratios of time intervals. It meets the standard of Euclidean rigor. An algebraic substitution of $\frac{x_A}{c}$ and $\frac{x_B}{c}$ respectively for T_A and T_B yields $\left(t_A - \frac{x_A}{c}\right) : \left(t_B - \frac{x_B}{c}\right) :: \left(t_B + \frac{x_B}{c}\right) : \left(t_A + \frac{x_A}{c}\right)$. This proportion defines the relations between intervals of space and time in two inertial reference frames moving relative to one another. Note that the only physical quantities represented are intervals of time, distances in space, and the speed of light. Moreover, only the speed of light is invariant (albeit not frame-independent).

By substituting in the proportion above the algebraic terms $\frac{x_A}{c}$ and $\frac{x_B}{c}$ we have implicitly introduced numerical units of time and distance and we must now make those units explicit. We do so by employing the law of cross-multiplication for proportions and writing a new proportion between compounded ratios: $\left(t_A - \frac{x_A}{c} : unit\right)\left(t_A + \frac{x_A}{c} : unit\right) :: \left(t_B - \frac{x_B}{c} : unit\right)\left(t_B + \frac{x_B}{c}\right) : unit$. We now translate the proportion into an algebraic equation by nominalizing the ratios as fractional numbers and reading the compounding of ratios as multiplication:

$$\left(t_A + \frac{x_A}{c}\right)\left(t_A - \frac{x_A}{c}\right) = \left(t_B + \frac{x_B}{c}\right)\left(t_B - \frac{x_B}{c}\right)$$

Multiplying through by c yields $(ct_A + x_A)(ct_A - x_A) = (ct_B + x_B)(ct_B - x_B)$. Finally, upon multiplying out the factors, we obtain the desired equation:

$$c^2t_A^2 - x_A^2 = c^2t_B^2 - x_B^2$$

Clearly, in the algebraic version of the Lorentz transformation, ratios or *relations* between quantities have been translated symbolically as quantities in their own right (fractional numbers). Subjected to the arithmetical operations of multiplication and division these ratios yield the squared terms of the Minkowski spacetime interval. That is to say, the squared terms in the spacetime interval are symbolic abbreviations for compounded ratios. They are not actual quantities. The algebraic expression for the spacetime interval thus arises through the same process of historically sedimented symbolic abbreviation as we saw in the historical examples discussed in Chap. 5.

Let us pause over this result. Were the so-called spacetime interval $c^2t^2 - x^2$ an actual physical or geometrical quantity, the squared terms would also have to be actual quantities. But the squared terms in the expression $c^2t^2 - x^2$ are no more actual quantities than is t^2 in the algebraic version of Galileo's law of fall. To be sure, for a given "space-like" interval we can always choose an inertial frame relative to which that interval reduces to the Pythagorean distance formula $s^2 = x^2 + y^2 + z^2$, in which case the squared terms of the interval indeed correspond to geometrical squares. But when we regard x^2 as the numerical measure of the area of a geometrical square, then we are applying the operation of multiplication improperly in Newton's sense, for "[the] generation of a surface by lines is very different from multiplication."[3] Instead, our warrant for adopting the custom of multiplying "length times width" to obtain area is underwritten by the Euclidean proof (Book VI, Prop. 23) that the area of a rectangle is proportional to the compounded ratio of its sides. In the case of a geometrical square, the compounded ratio of interest is a duplicate ratio and so the proportion translates in numerical terms to Area = length × width. But all that shows is that I am free to write the Pythagorean Theorem algebraically as $s^2 = x^2 + y^2 + z^2$. Clearly, I may not assume the converse and assert that the algebraic term x^2 in the so-called spacetime interval represents a geometrical square. For it may or may not represent one depending on whether it is geometrical in origin. The expression for the space-like interval $x^2 + y^2 + z^2$ has its origin in the Lorentz transformation, not in geometry, and so we merely recover the same numerical values as the Pythagorean distance formula. It is not on its own terms a geometrical formula.

The theory of four-dimensional spacetime depends for its physical validity on our understanding the interval $c^2t^2 - x^2$ as a real quantity in nature. However, through our historical desedimentation the interval reveals itself as an algebraic artifact rather than a physical or geometrical quantity. I submit this result as conclusive: *The theory of Minkowski spacetime mistakes a symbolic construction for a physical reality.* Minkowski spacetime may be a useful or even a beautiful fiction, but it is a fiction nonetheless, for the theory of spacetime fails precisely on the point of its *sine qua non*: the interval as a real physical quantity.

We may continue to call the relativistic manifold of events a "continuum," as long as we recognize that it is not a single metrical continuum of merged space and time. To repeat, the metrical properties of space and time in the theory of relativity are interdependent or "entangled," as per Einstein 1905, but they are not merged into a single geometrical continuum of four dimensions. We shall now see that the same holds for general relativity.

Notes

1. Einstein 1961 [1916], 62–63.
2. I have appropriated the example from Bondi (1964), where it is treated algebraically. This particular analysis of the scenario originally appeared in Cosgrove 2012, 175–176.
3. Newton 1769 [1707], 11.

PART III

General Relativity Without Spacetime

CHAPTER 7

Minkowski Spacetime and General Relativity

Einstein famously quipped in his popular book on relativity that the general theory of relativity "would not have gotten out of diapers" apart from Minkowski's four-dimensional formalism. We must now show that general relativity can be formulated apart from the concept of Minkowski spacetime. To this purpose, however, it is not sufficient to demonstrate that general relativity relies merely on Minkowski's mathematical formalism, without committing itself to the physical reality of Minkowski spacetime. For the question remains as to the basis for the heuristic value of the mathematical formalism, if it be not the physical reality of Minkowski spacetime itself. The indispensability of Minkowski's formalism, on the regnant interpretation, lies in its providing the requisite quadratic differential form for employment of Riemannian geometry. But in itself the employment of the mathematics of Riemannian geometry hardly renders general relativity "geometrical" in subject matter, as Einstein was well aware. Nevertheless, it is not obvious why Riemannian geometry would be effective in general relativity unless the Minkowski line element were physically real, since the metrical structure of any Riemannian manifold is defined by its line element.

Our task, then, is to demonstrate both that Minkowski spacetime as a *physical concept* is superfluous to general relativity and that the whole mathematical apparatus of four-vectors is superfluous as well. That means we must address the two respects in which Minkowski's theory enters standard formulations of the general theory of relativity: the field equation

J. K. Cosgrove, *Relativity without Spacetime*,
https://doi.org/10.1007/978-3-319-72631-1_7

itself and the law of geodesic motion. In some quarters, of course, there is a preference for deriving the law of motion directly from the field equation.

7.1 Tensor Calculus and "Geometrical Objects"

If you look up the definition of "tensor" or the "tensor calculus" in physics textbooks or other mathematical sources you will most likely learn that a tensor is a "geometrical object" that "maps vectors to real numbers" or some such thing. The more traditional view of a tensor as a purely analytical entity, defined in terms of its transformation properties, appears to have been more or less supplanted by the geometrical definition at least since the nineteen-sixties or thereabouts. The geometrical definition unfortunately is misleading, not least in its suggestion that tensors, by virtue of their invariance properties, are what is "really there" in nature.

7.1.1 Tensors as Ratio-Compounding Machines

Should tensors be defined as "geometrical objects"? A little reflection suggests that they should not. To be sure, tensors sometimes are used to represent geometrical quantities or subject matter (for example, the Riemann tensor in differential geometry); but sometimes they are used to represent physical quantities (for instance, the stress tensor of classical physics) and sometimes they represent purely numerical relationships (the metric tensor, for instance, which determines the numerical coefficients of a quadratic differential function). In the specific case of the quadratic differential form employed by Einstein for the metric of general relativity ($\varphi = a_{rs}dx_r dx_s$), John Norton has reminded us that,

> [o]f course the modern reader immediately associates this form with the invariant line element of a non-Euclidean surface of variable curvature, such as was introduced by Gauss and developed by Riemann. However Ricci and Levi-Civita's $x_1, \ldots, x_n$ were *variables* and not necessarily geometrical coordinates. They were at pains to emphasize that what was then called infinitesimal geometry was just one of many possible applications of their calculus.[1]

Tensors by definition are *analytical entities*, then, not geometrical entities. For it surely makes no sense to define something in terms of merely one of the possible instances falling under that definition. Indeed, the best way to define a tensor is the traditional way in terms of its transformation properties. For instance, we define a covariant tensor of first rank in analytical terms as a set of components which transform according to $A'_\alpha = \left(\frac{\partial x_\gamma}{\partial x'_\alpha}\right) A_\gamma$. And the preceding can be suitably generalized to cover tensors of any rank/type.

Why do standard definitions so often call tensors "geometrical objects"? Speaking in terms of a vector or tensor of first rank, we observed in Chap. 2 that a vector in the proper sense of the term is a physical or geometrical quantity with a direction in space. However, since vectors can be resolved into directional components subject to a transformation rule, any set of numbers that so transform may be regarded as a "vector" in the purely analytical sense, even if the "components" in question are not components of any single directed physical or geometrical quantity. So in this purely analytical sense we indeed may say that a tensor maps vectors to real numbers, and sometimes it really does (the metric tensor of differential geometry, for instance). But this sense of vector in no wise necessitates our regarding tensors as geometrical objects, for the vectors so mapped are not necessarily themselves geometrical objects.

What the tensor calculus truly is, in light of Vieta's dictum that an equation resolves a proportion, is a machine for compounding ratios. We can support this definition with some simple observations on the transformation properties of a tensor. Consider once again the coordinate transformation for a rotation in Euclidean space:

$$x' = x cos\theta - y sin\theta$$

$$y' = x sin\theta + y cos\theta$$

The transformation above is obtained by summing the unprimed components of each component in the primed system. Each of these unprimed components is obtained in turn from the ratio of the change in the primed component to a unit change in the unprimed component—for example, $\Delta x : \Delta x' :: \cos\theta : 1$, which we can translate into algebra as $\partial x'/\partial x = cos\,\theta$. So $\partial x'/\partial x$ is an algebraically nominalized ratio of the kind we have discussed, a symbolic abbreviation in d'Alembert's sense. If we then regard x'_1 and x'_2 as components of vector A', we can write $A'^\gamma = (\partial x'_\gamma/\partial x_\delta)A^\delta$, always remembering that the $\partial x'_\gamma/\partial x_\delta$ are symbolically abbreviated ratios. But $A'^\gamma = (\partial x'_\gamma/\partial x_\delta)A^\delta$ is a contravariant tensor of rank one. And so a contravariant tensor of rank two is obtained by compounding two algebraically nominalized ratios:

$$A'^{\gamma\rho} = \left(\partial x'_{\gamma} / \partial x_{\delta}\right)\left(\partial x'_{\rho} / \partial x_{\sigma}\right) A^{\delta\sigma},$$

And so forth for tensors of higher rank. For each higher rank, we stack on an additional compounded ratio. That is what I meant above by saying that the tensor calculus is a ratio-compounding machine.

The advantage of Cartesian algebra over the traditional mathematics of ratio and proportion, however, extends well beyond the former's superior calculational power. Algebra also enhances the representation of quantitative relationships. This conceptual advantage lies precisely in the nominalizing of ratios as quantities, which allows for the free substitution of algebraic quantities. Recall that in the process of constructing a tensor above we had to rewrite the ratios as numerical fractions, which alone allowed us to apply the ratios to the transformed vector components. Similarly, to derive the Minkowski interval in by means of ratio and proportion in Chap. 6, we at a certain point had to substitute $\frac{x}{c}$ for T in the proportion $(t_A - T_A) : (t_B - T_B) :: (t_B + T_B)(t_A + T_A)$. We were licensed to make this substitution because the ratio of velocities is directly proportional to the ratio of distances traversed and inversely proportional to the ratio of times elapsed, which translates into algebraic as $c = \frac{x}{t}$. Without the algebraic substitution we never could have incorporated the spatial terms and thus would have been brought to a stop. We are therefore far from recommending a return to traditional Euclidean ratio and proportion as the primary language of mathematical physics. However, the power of modern symbolic algebra comes at the price of a loss of meaning. It is this inevitable loss that Newton reprehended in his polemics against Cartesian algebra.

7.1.2 Tensors and Invariance

A tensor is a symbolic construction and therefore cannot be what is "really there" at a point in space and time. It is sometimes, but not always, the case that a tensor represents some frame independent physical quantity that is really there. The classical stress tensor, for instance, returns a single invariant and frame independent physical quantity, the total stress at a point. The metric tensor of differential geometry, on the other hand, itself represents no geometrical quantity but is rather a matrix of numerical coefficients. It is only when this matrix "acts upon" the appropriate quadratic differential form that a single invariant geometrical quantity (ds^2 or the square of the distance) is recovered. But the tensor is the numerical matrix itself, which neither is an invariant geometrical object nor is "really there."

In fact, it is not true that a tensor must be even *associated* with any single invariant quantity. With respect to the electromagnetic field tensor in special relativity, for instance, there is no such single physical invariant. Depending on the inertial frame of reference we divide up the magnetic and electrical components of the electromagnetic field differently, but there is determined no single physical quantity over and above the set of tensor components. After all, the electric field has different units than the magnetic field, so we hardly could combine them in a single physical quantity. The only sense in which we can employ a so-called "component-free" approach to the electromagnetic field, therefore, is by means of the purely notational gesture of writing $\boldsymbol{F}$ instead of F^{ab}. It is fundamentally misleading to call $\boldsymbol{F}$ an "intrinsic" representation, as if $\boldsymbol{F}$ directly represents what is really there, while F^{ab} goes indirectly by means of components. The electromagnetic field always has electric and magnetic components—that has nothing to do with coordinate representation and is a fact about the physical world.

With respect to the metric tensor of the general theory of relativity, to be sure we can represent the four coordinate differentials of interest (three spatial and one temporal) as components of a four-dimensional displacement vector and then "map vectors to real numbers" by means of the metric tensor. But clearly in this case we are dealing with a "displacement vector" strictly in the analytical or symbolic sense, not in the proper sense of a vector as a single physical or geometrical quantity with direction space. Since the time component is not spatial, the four-dimensional entity of which we call it a "component" cannot be a vector in the physical or geometrical sense. Thus the mere use of tensors in the formulation of the general theory of relativity in no wise implies the theory itself is geometrical or that it deals exclusively with geometrical invariants.

7.2 Against the Long Clothes: Minkowski Spacetime in Einstein's 1916 Review Article

Although Einstein did indeed avail himself of Minkowski's four-vector formalism in the general theory of relativity, he need not have and the concept of Minkowski spacetime itself plays no substantive role in the theory.

Einstein was wrong to suggest that the theory could not have been formulated without it. The best way to demonstrate the point is to carefully review Einstein's 1916 paper on general relativity, his first comprehensive presentation of the completed theory. The paper consists of five parts: Part A outlines the conceptual structure of the theory, Part B sets forth the required mathematical apparatus for formulating the theory, Part C sets forth the gravitational field law, Part D applies the theory of gravity to special relativistic physics, and Part E derives Newton's theory of gravity as a first approximation. We shall be interested in Parts A, B, C, and E.

7.2.1 Einstein 1916, Part A: "Fundamental Considerations on the Postulate of Relativity"

Einstein launches his paper with the observation that the special theory of relativity, for all of its far-reaching modifications to classical mechanics, nevertheless retains two of the latter theory's key assumptions: first, the principle of relativity is limited to inertial frames of reference, and second, the laws of kinematics pertain to measurements by means of rigid bodies (rods and clocks) which maintain their metrical properties independently of place and time. Einstein proceeds in section 2 of Part A to offer two independent arguments for an extension of the principle of relativity to accelerated frames of reference. The first argument is epistemological ("from the theory of knowledge") and essentially derived from "Mach's principle." Since Einstein later renounced it we can leave it out of consideration for present purposes. The second argument for the extension of the principle of relativity appeals to Einstein's principle of equivalence.

7.2.1.1 Part A, §2: The Principle of Equivalence

The theory of gravity is first introduced in Einstein's paper as a necessary condition for a generalized principle of relativity: "It will be seen from these reflections that in pursuing the general theory of relativity we shall be led to a theory of gravitation, since we are able to 'produce' a gravitational field merely by changing the system of coordinates" (114).[2] Here I shall emphasize *Einstein's* principle of equivalence because the term has taken on a sense in the literature which departs from the sense in which Einstein employed it.[3] The principle of equivalence in Einstein's version, which is in my judgment the only one coherent with the conceptual structure of general relativity, states that for a finite region in which the special theory of relativity holds, an accelerated reference frame induces a special kind of gravitational field. The requirement for a *finite* special relativistic region is clear from Einstein's application of the principle of equivalence in his derivation of the metric geodesic later in the 1916 paper: "We now make the assumption, which readily suggests itself, that this covariant system of equations also defines the motion of the point in a gravitational field in the case when there is no system of reference K_0, with respect to which the special theory of relativity holds good in a finite region."[4] The principle of equivalence thus not only extends the principle of relativity to non-inertial reference frames in finite special relativistic regions, thus removing an epistemological defect of special relativity (the absolute status of the inertial frame), but also accounts for the equivalence of inertial and gravitational mass and forges the needed link between special relativity and the theory of gravity.

Since for a finite region in which the special theory of relativity holds ("Galilean region") a free body follows an inertial path, we can deduce how that body will move relative to an accelerated frame in the same Galilean region. This accelerated motion is to be regarded as proceeding under the influence of a gravitational field of a special kind not associated with source masses. Then, on the basis of a reasonable conjecture, we can apply the generally covariant equation of geodesic motion to the case of a gravitational field for which special relativity does not hold good in a finite region. The preceding line of reasoning can be sustained, however, only if the principle of equivalence is not infinitesimal and the gravitational fields underwritten by the principle of equivalence are not fictitious or merely apparent. Einstein explains as much to Laue in response to the latter's suggestion that, in view of the vanishing of the Riemann tensor in the rigidly rotating disk scenario, the gravitational field associated with the rotating disk should be regarded as fictitious:

> What characterizes the existence of a gravitational field from the empirical standpoint is the non-vanishing of the Γ^l_{ik}, not the non-vanishing of R_{iklm}. If one does not think intuitively in such a way, one cannot grasp why something like a curvature should have anything at all to do with gravitation. In any case, no reasonable person would have hit upon such a thing. The key for understanding the equality of inertial and gravitational mass is missing.[5]

Thus, even though the Riemann tensor vanishes for the gravitational field of the rotating disk—since a finite special relativistic region is presupposed by the thought experiment—the Γ^l_{ik} or affine connection still registers the real presence of a gravitational field.

The preceding observations are difficult to reconcile with a vital role for Minkowski spacetime in the general theory of relativity. If we embrace Einstein's version of the principle of equivalence, according to which the accelerated motion of a reference body relative to a finite Galilean region induces a real gravitation field, then the concept of the gravitational field as a frame independent geometrical object ceases to make sense. For on the relativistic view sanctioned by the principle of equivalence, the inertio-gravitational field must break down into inertial and gravitational components depending on the relative state of motion of the reference system. And as we are so often reminded by expositors of Minkowski's theory, "nothing moves in spacetime." It is of no avail in this context to object that in Minkowski spacetime, frame relative phenomena are determined by different "slicings" of the absolute four-dimensional geometrical continuum which alone is real. For Einstein's principle of equivalence requires that relative motion itself be real. As Synge saw so clearly, an absolute geometrical formulation of Einstein's theory of gravity cannot possibly countenance any such role for the motion of a reference frame as the principle of equivalence assigns it:

> ... the geometrical way of looking at space-time comes directly from Minkowski. He protested against the use of the word "relativity" to describe a theory based on an "absolute" (space-time), and, had he lived to see the general theory of relativity, I believe he would have repeated his protest in even stronger terms.... I have never been able to understand this Principle [of Equivalence] ... In Einstein's theory, either there is a gravitational field or there is none according to whether the Riemann tensor vanishes. This is an absolute property; it has nothing to do with any observer's world-line.[6]

Obviously Einstein himself would not have agreed with respect to the application of the principle of equivalence in finite Galilean regions, and we have taken Einstein's side on the matter. Nevertheless, Synge is certainly correct that the principle of equivalence is incompatible with Minkowski's theory. And if, against Einstein's interpretation, we instead regard the principle of equivalence as infinitesimal we likewise sever the conceptual link between special and general relativity and with it any possible connection between Minkowski spacetime and the general theory of relativity.

According to John Stachel, however, it is precisely with respect to the considerations raised by the rotating disk application of the principle of equivalence that Minkowski's theory comes into play. For the hypothesis of a non-flat metric to space and time led Einstein to conclude that "what was needed was a four-dimensional generalization of Gauss's two-dimensional surface theory, and that the flat metric tensor of Minkowski's formulation of special relativity had to be generalized to a non-flat metric".[7] However, what Stachel here refers to as the flat metric tensor of Minkowski's formulation is simply the coefficients −1, −1, −1, and 1 attached to the variables dx_1^2, dx_2^2, dx_3^2, and dx_4^2 respectively, in the Lorentz transformation $-dx_1'^2 - dx_2'^2 - dx_3'^2 + dx_4'^2 = -dx_1^2 - dx_2^2 - dx_3^2 + dx_4^2$. And not only is Minkowski's formulation not required for the requisite transformation of variables, but it arguably strips the expression $-dx_1^2 - dx_2^2 - dx_3^2 + dx_4^2$ of physical intelligibility by investing it with a physical meaning independently of its role in the transformation equation from which it is derived.

No doubt Einstein was thinking along the lines Stachel describes. But our present interest is not whether Einstein used Minkowski, as he obviously did, but whether he had to. In fact, the rotating disk experiment itself involves no merging of space and time into a single metrical continuum as per Minkowski's theory. Rather, we derive the non-Euclidean metrical properties of *space* on the disk via special relativistic length contraction of the small measuring rods around the circumference of the disk, and we derive the metrical properties of *time* on the disk via retardation of clocks situated at the circumference compared with the center of the disk. That is to say, the two metrical effects of interest, spatial and temporal, are independently derived, with no merging of space and time and no necessary reference to a geometrical line element.

John Norton proposes what at first glance seems a more weighty argument on behalf of Minkowski's four-dimensional spacetime formulation. Once we relinquish rigid frames, Norton observes, we are deprived of the three-dimensional relative spaces that Einstein has used thus far to articulate the conceptual structure of the theory:

> It was inevitable that Einstein would give up the use of standard [three-dimensional] formulations of theories in his search for a general theory of relativity. For the relative spaces used by these formulations would only have well-defined geometries if the associated frame is in rigid motion, which is by no means generally the case. Even in Minkowski space-time, no *non*uniformly rotating frame can move rigidly. Worse, the relative space will only have the frame time required by the standard formulation if the space-time admits a foliation by hypersurfaces orthogonal to the frame. Even uniformly rotating frames in Minkowski space-time do not admit such a foliation. (Norton 1989, 25)

Einstein was thus compelled by the breakdown of rigid frames, Norton argues, to embrace a four-dimensional spacetime formulation for his theory of gravity. The rotating disk thought experiment, itself an application of the principle of equivalence in a three-dimensional relative space, thus ultimately leads to the demise of three-dimensional relative spaces. And the only alternative is some generalization of Minkowski spacetime.

Einstein's own conclusion from the rotating disk thought experiment is rather that a generally covariant formulation must be employed, since the usual method for assigning coordinates by means of rigid rods and clocks has broken down on the disk. There is no mention of Minkowski spacetime. In the event, the notorious "reference mollusk" of Einstein's popular book is somewhat more informative on this point. Einstein observes that by continuing to speak of "reference bodies" in the context of the theory of gravity we are making a concession to "our 'old-time' three-dimensional view of things." Nevertheless, if we insist on that view of things, we can perhaps speak of "non-rigid reference bodies … which are as a whole not only moving in any way whatsoever, but which also suffer alterations in form *ad lib.* during their motion."[8] The only reason to enlist the reference mollusk in preference to formal Gaussian coordinates, Einstein notes, is that the former allows for "the (really unjustified) formal retention of the separate existence of the space coordinates as opposed to the time coordinate." Hence the demise of three-dimensionality indeed seems to be complete once we are deprived of rigid frames.

That can hardly be the case, though, for what could Einstein mean by asserting in the passage above that the "separate existence" of the space coordinates and time coordinate is merely formal and even "unjustified"? The claim in fact refers back to a version of the so-called point-coincidence argument from the preceding chapter of the same book:

> We refer the four-dimensional space-time continuum in an arbitrary manner to Gauss coordinates. We assign to every point of the continuum (event) four numbers, x_1, x_2, x_3, x_4 (coordinates), which have not the least direct physical significance, but only serve the purpose of numbering the points of the continuum in a definite but arbitrary manner. This arrangement does not even need to be of such a kind that we must regard x_1, x_2, x_3 as "space" coordinates and x_4 as the "time coordinate."[9]

By numbering points in a "definite but arbitrary manner" Einstein means assigning coordinates within the minimal formal requirements of a differentiable manifold; and by the assertion that we do not even need to distinguish which are the space coordinates and which the time coordinate he evidently means that, once the assignments are complete, we can manipulate the variables mathematically without regard to which of them represent time and which represent space. From a physical point of view, of course, we must distinguish the time coordinate from the space coordinates, if only because the time coordinate enters the equations with a different sign. Einstein's faith in the point-coincidence argument seems somewhat overweening in any event, since the purpose of a coordinate system is not simply to assign unambiguous *labels* to point-events, any more than the aims of a physical theory can be reduced to merely cataloging the careers and intersections of the world-lines associated with such point events. All we really learn from the point-coincidence argument is that since, in general, we can no longer utilize rigid reference bodies in gravitational fields, we do well to embrace Gaussian coordinates.

But is the embrace of Gaussian coordinates in the general theory of relativity tantamount to the embrace of Minkowski's four-dimensional spacetime? At this point in Einstein's discussion, at least, that conclusion would follow only if the demise of the three-dimensional relative spaces of which Norton speaks implied the demise of three-dimensional relative spaces as such. But it clearly does not. The three-dimensional relative spaces suffering demise in Norton's account, after all, are three-dimensional relative spaces with well-defined geometries and frame times. These well-defined geometries and frame times are necessary for the principle of equivalence to serve its purpose by enabling us to deduce the properties of gravitational fields in general by means of the properties of special fields in finite Galilean regions. Already with the rigidly rotating disk, however, we have been deprived of a frame time. Does that compel us to set aside the three-dimensional interpretation of the rotating disk's gravitational field in favor of a Minkowskian four-dimensional space-time account? In no wise. That would defeat the very logic of the principle of equivalence and render the gravitational field of the disk fictitious, since the field is induced by the *relative motion* of the disk and inhabits the three-dimensional relative space of the disk. To be sure, for gravitational fields more generally the principle of equivalence has already served its purpose and we can no longer employ rigid reference bodies. But none of this has anything to do with Minkowski spacetime.

We conclude that the demise of three-dimensional relative spaces in Norton's argument solely regards such relative spaces as are associated with rigid reference bodies, not three-dimensional relative spaces per se.

Hence, the demise of rigid frames in no way compels us to embrace Minkowski four-dimensional spacetime in the formulation of the general theory of relativity. For Einstein himself, it rather compels the embrace of Gaussian coordinates. But we have yet to be given a reason why the employment of Gaussian coordinates and the associated mathematical machinery (Riemannian geometry and the tensor calculus) is in any way related to the embrace of Minkowski spacetime.

7.2.1.2 Part A, §3: General Covariance

Einstein proceeds in Part A, section 3 to offer two arguments on behalf of general covariance. Curiously, he makes no direct inference from the general principle of relativity to general covariance, but rather appeals in the first place to the unavailability of rigid reference frames in a gravitational field (rotating disk argument) and then to the point-coincidence argument for the arbitrariness of coordinate assignments. Neither of these arguments actually establishes a requirement for general covariance. What Einstein actually demonstrates by means of the rotating disk is that in the presence of a gravitational field we lose the direct metrical significance of coordinate differentials. But that entails nothing more than that some alternative to rigid Cartesian frames is required, so we can represent all possible gravitational fields countenanced by the theory. It does not necessitate general covariance. General covariance provides for differing coordinate representations of the *same* gravitational field, that is, while the demise of rigid frames entails rather our finding a means of coordinate representation for *different* gravitational fields. The ill-fated *Entwurf* theory, for instance, which employed a generally covariant metric tensor along with a field law that was not generally covariant, was ultimately rejected by Einstein not on account of its restricted covariance but for quite other reasons.[10]

Neither does the point-coincidence argument establish a requirement for general covariance as opposed to merely an option:

> The introduction of a system of reference serves no other purpose than to facilitate the description of the totality of such coincidences [intersections of material point-events]. We allot to the universe four space-time variables x_1, x_2, x_3, x_4 in such a way that for every point-event there is a corresponding system of values of the variables $x_1 \ldots x_4$.... As all our physical experience can be ultimately reduced to such coincidences, there is no immediate reason for preferring certain systems of coordinates to others, that is to say, we arrive at the requirement of general covariance. (117–118)

Since coordinates merely serve the purpose of labeling point-events, so long as that is accomplished unambiguously no particular coordinate system or set of coordinate systems is to be preferred over any other. But such lack of favoritism by nature does not entail any requirement for general covariance and, as has often been noted in the literature, it would apply as well to any theory, not just general relativity.

Since both the rotating disk and point-coincidence arguments for general covariance fail, it is strange that Einstein did not offer a direct argument for general covariance based on the general principle of relativity itself. While the rotating disk argument does make an indirect appeal to the general principle of relativity, by way of the principle of equivalence, its proximate basis is the breakdown of rigid coordinate systems. Clearly Einstein distinguished general covariance per se from general relativity, for at the close of his discussion of the rotating disk he adverts to the fact that the general covariance group of transformations is larger than the general relativity group:

> It is clear that a physical theory which satisfies this postulate [general covariance] will also be suitable for the general postulate of relativity. For the sum of *all* substitutions in any case includes those which correspond to all relative motions of three-dimensional systems of coordinates. (117)

In the 1916 paper, however, Einstein seems to have regarded general covariance as a *sufficient* condition for general relativity, whereas after Kretschmann's objection he was forced to admit this was not the case. What we can say from our own vantage point is that *extended* covariance—or covariance beyond Lorentz covariance—but not general covariance per se is a *necessary* condition for the realization of a general principle of relativity. The requisite degree of extended covariance is that which accommodates "all relative motions of three-dimensional systems of coordinates" in finite Galilean regions, as per the principle of equivalence. This includes, for instance, the rotating frames of reference upon which the *Entwurf* theory finally foundered in 1915.

The topic of general covariance would hardly be relevant to our topic were it not for the related thesis, which has gained considerable currency over the last decades, that general covariance is a minimal requirement for any "well-formulated theory." While the precise sense of the concept "well-formulated" is not always clear in the literature, the basic claim evidently is that theories formulated "intrinsically" in terms of component-free geometrical objects are trivially generally covariant, since coordinates are in that case optional.[11] But the terms "coordinate-free" and "component-free" clearly must be distinguished: a vector, for instance, can be resolved into components without the use of a coordinate system, as Galileo did with the horizontal and vertical components of ballistic trajectories in space and Newton with his parallelogram of forces. With specific regard to components, the assertion that general relativity is "well-formulated" only if it is formulated "intrinsically" clearly begs the question, for what is at issue is precisely whether the theory must or should be formulated *geometrically* as per Minkowski's theory. It is indeed true that in geometry not only are coordinates optional, but also components. However, even in the context of Minkowski's theory a "component-free" representation can only be a notational gesture, since Minkowski's four-vectors are originally constructed out of components. At a minimum, and even if we represent the spatial component intrinsically without resolving it into the x, y, and z components, the time component must always be distinguished.

7.2.1.3 Part A, §4: The "Linear Element"

Einstein explicitly acknowledges Minkowski, in the introduction to his 1916 review article, as having been the "first one to recognize the formal equivalence of space coordinates and the time coordinate."[12] The concept of Minkowski spacetime itself, however, first makes its appearance in section 4 of the paper, where Einstein notes that on the assumption that special relativity holds for infinitesimally small regions of space and time,

> the expression $ds^2 = -dx_1^2 - dx_2^2 - dx_3^2 + dx_4^2$ then has a value which is independent of the orientation of the local coordinates, and is ascertainable by measurements of space and time. The magnitude of the linear element pertaining to points in infinite proximity, we call *ds*....
>
> To the "linear element" in question, or to the two infinitely proximate point-events, there will also correspond definite differentials dx_1 ... dx_4 of the four-dimensional coordinates of any chosen system of reference. (Einstein 1952b [1916], 119)

The quotation marks around "linear element" in its second occurrence above perhaps are intended simply to introduce the reader to the new and potentially unfamiliar concept of the spacetime interval, but Einstein could also be signaling that he means to employ the concept of "linear element" in a strictly formal sense, in line with his earlier comment that Minkowski's significance lies in his having noticed the formal equivalence of the space and time coordinates.[13] The decisive question is this: Does any concept Einstein introduces in the entirety of the paper necessitate our regarding the quadratic differential expression $-dx_1^2 - dx_2^2 - dx_3^2 + dx_4^2$ as a line element in spacetime or indeed as anything related to Minkowski's theory at all? For on its face the expression is merely a Lorentz invariant as in special relativity 1905, and so we might well dispense with ds^2 and write rather $-dx'^2_1 - dx'^2_2 - dx'^2_3 + dx'^2_4 = -dx_1^2 - dx_2^2 - dx_3^2 + dx_4^2$, with no reference to Minkowski's theory.

Einstein explains in a footnote that he has chosen a unit of time to make the velocity of light unity, but we should note that he also suppresses the units of the velocity of light by writing the time coordinate as simply x_4. This is a harmless notational gesture as long as we remember that the units are still there and that the time component of the so-called linear element carries units of spatial distance (ct). Einstein next introduces the familiar quadratic differential equation $ds^2 = g_{\sigma\tau} dx_\sigma dx_\tau$ that we associate with Riemannian geometry.[14] Now if it could be demonstrated that the preceding equation, as it functions in general relativity, is actually derived from Riemannian geometry we would have our first concrete evidence that the general theory of relativity indeed depends upon a geometrical formulation based on the concept of Minkowski spacetime; for the employment of Riemannian geometry indeed requires a line element. However, Einstein's actual derivation of the equation disappoints that hope since it has nothing to do with Riemannian geometry. He rather takes the Lorentz transformation $-dx'^2_1 - dx'^2_2 - dx'^2_3 + dx'^2_4 = -dx_1^2 - dx_2^2 - dx_3^2 + dx_4^2$ and substitutes generalized coefficients for the transformed variables with the restriction to inertial frames removed. That is to say, for the x coordinate Einstein obtains $x' = Ax + By + Cz + Dt$, for the time coordinate $t' = Ex + Fy + Gz + Ht$, and similarly for the other coordinates. In Einstein's more compact notation the transformations are thus represented as $dX_\nu = a_{\nu\sigma} dx_\sigma$, where $a_{\nu\sigma}$ designates collectively our coefficients A, B, C, …, and so forth. Substitution of the transformed variables into the Lorentz equation $-dx'^2_1 - dx'^2_2 - dx'^2_3 + dx'^2_4 = -dx_1^2 - dx_2^2 - dx_3^2 + dx_4^2$, again derived from special relativity 1905 with no relation to Minkowski, yields $g'_{\sigma\tau} dx'_\sigma dx'_\tau = g_{\sigma\tau} dx_\sigma dx_\tau$. Einstein of course instead writes $ds^2 = g_{\sigma\tau} dx_\sigma dx_\tau$, but that is merely a notational gesture at this point, since he has not yet demonstrated that we are dealing with any such thing as a line element or "spacetime interval." There is no justification as of yet for writing ds^2 on the left-hand side. That is to say, the resemblance of the expression to the line element of Riemannian geometry alerts us to a formal analogy *and no more*, which would be obvious if we wrote out all the terms in such a way as to the show the opposing signs of the space and time variables. Nothing of substance is accomplished by coining the term "semi-Riemannian" to account for the difference in metric signature. Prima facie, then, the relativistic equation $g'_{\sigma\tau}dx'_{\sigma}dx'_{\tau} = g_{\sigma\tau}dx_{\sigma}dx_{\tau}$, derived by generalizing the Lorentz transformation, represents nothing geometrical and is not an application of Minkowski's theory.

Einstein concludes section 4 with the observation that, in accord with his earlier application of the principle of equivalence, the coefficients $g_{\sigma\tau}$ are to be regarded as quantities describing the gravitational field relative to a given reference system (the formal equivalent, that is, of the scalar gravitational potential in Newton's theory). Thus, if in a Galilean region a free material point moves uniformly in a straight line, such that the $g_{\sigma\tau}$ assume constant diagonal values −1, −1, −1, and +1 relative to a suitably chosen frame of reference, then relative to an arbitrarily accelerated frame the $g_{\sigma\tau}$ will assume non-constant values as a function of the space and time coordinates. Furthermore, that same material point will now move non-uniformly, with the law describing its motion being independent of the nature of the moving particle. It follows that the particle's non-uniform motion is to be regarded as occurring under the influence of a gravitational field described by the $g_{\sigma\tau}$. Moreover, and this is the decisive move in Einstein's argument, even "in the general case, when we are no longer able by a suitable choice of coordinates to apply the special theory of relativity to a finite region, we shall hold fast to the view that the $g_{\sigma\tau}$ describe the gravitational field" (120).

Einstein has by this point in his paper (sections 1–4) laid out the essential conceptual structure of the general theory of relativity, with no discernible role for Minkowski's theory. All we have with respect to Minkowski spacetime, in fact, is the physically unjustified phrase "linear element" and the gratuitous notational gesture "ds^2." And if by this juncture Minkowski's theory has not assumed a physically substantive role in Einstein's theory, one could be forgiven for wondering at what point it reasonably would be expected to fulfill such a role. After all, Einstein forthwith proceeds in Part B to what he calls the "purely mathematical task" of finding a generally covariant field equation in terms of the calculus of tensors, and if the task is purely mathematical we should not expect to encounter any new physical concepts. The remaining candidates for a substantive role for Minkowski's theory, therefore, would appear to be

(1) the law of geodesic motion, usually understood in terms of a "straight line in four-dimensional spacetime," and (2) the stress-energy tensor, a generalization of the Minkowski four-momentum.

7.2.2 *Einstein 1916, Part B: "Mathematical Aids to the Formulation of Generally Covariant Equations"*

We can skip over much of sections 5–12 in Part B on the mechanics of the tensor calculus. Of interest, however, is section 8 on the "fundamental tensor," section 9 on the equation for a geodesic, and Einstein's introduction of the Riemann tensor in section 12. Einstein treats tensors as purely analytical entities in these mathematical sections, in accord with our account in Sect. 7.1 above. He elaborates helpfully in his Princeton lectures, explaining how we can define a vector in analytical terms without reference to geometry per se:

> The ensemble of three quantities, defined for every system of Cartesian coordinates, and which transform as the components of an interval, is called a vector. If the three components of a vector vanish for one system of Cartesian coordinates, they vanish for all systems, because the equations of transformation are homogeneous. We can thus get the meaning of the concept of a vector without referring to a geometrical representation.[15]

What defines a vector in the passage above is that the three quantities *transform as the components of an interval.* But the fact that the quantities in question transform in the same way as the components of a geometrical interval does not mean that they actually are components of an interval or anything geometrical at all. And the same applies to tensors: since a vector taken in the purely analytical sense is a tensor of rank one, "tensors of higher rank … may be defined analytically."[16]

7.2.2.1 *Part B, §8: Fundamental Tensor $g_{\mu\nu}$*

Einstein points out that, based on the mathematical proof he has given in section §7, the matrix $g_{\mu\nu}$ ("fundamental tensor") in the expression for the linear element $ds^2 = g_{\mu\nu} dx_\mu dx_\nu$ is a covariant tensor of second rank. As we observed above, though, the application of the concept "linear element" to this quadratic differential expression is physically gratuitous, since Einstein has derived the expression not from differential geometry but rather by means of substituting the transformed variables $dX_\nu = a_{\nu\sigma} dx_\sigma$ into the Lorentz equation $-dx_1'^2 - dx_2'^2 - dx_3'^2 + dx_4'^2 = -dx_1^2 - dx_2^2 - dx_3^2 + dx_4^2$.

Consequently, if we mean to call $g_{\mu\nu}$ a *metric tensor*, as we should, we as yet have no license for assuming that it is a metric tensor in the sense of differential geometry, where the $g_{\mu\nu}$ map the contravariant vectors dx_μ and dx_ν to a single real number, the square of the line element at a point. That is to say, it is altogether possible, and I shall argue that it is indeed the case, that the $g_{\mu\nu}$ determine the metrical properties of space and time at a point without doing so by means of a "spacetime line element." All we can say at this stage of Einstein's discussion is that $g_{\mu\nu}$ is a covariant tensor of second rank and that the metrical properties of the space and time in a gravitational field are to be represented by this particular analytical entity. Or, to put a finer point on the matter, although we do know that $g_{\mu\nu}$ will represent the geometrical properties of *space* in Einstein's theory, we do not know nor do we have any reason to expect that $g_{\mu\nu}$ will represent the metrical properties of *time* geometrically, or represent space and time together geometrically; for we have been given no reason to believe that the metrical properties of time are *geometrical*. Clearly, the fact that the metrical properties of both space and time can be represented in a single quadratic differential expression ($g_{\mu\nu}dx_\mu dx_\nu$) in no wise forces upon us the conclusion that space and time themselves are therefore merged into a single geometrical continuum. For we already know from Einstein's 1905 special relativity, apart from Minkowski, that the metrical properties of space and time are entangled and therefore must be represented in a single mathematical expression (the Lorentz transformation). There is nothing new on this count in general relativity.

7.2.2.2 Part B, §9: Mathematical Derivation of Geodesic Line

In his 1916 paper Einstein employs for the derivation of the geodesic line the metrical notion of stationary ds ($\delta \int ds = 0$) rather than tangent vector parallel transport. In 1917, Levi-Civita introduced the concept of parallel transport, derived from the metric, and subsequently Hermann Weyl in 1923 demonstrated that parallelism can in fact be defined apart from the metric. Einstein's use of stationary ds has suggested to some interpreters that in 1916 Einstein had yet to conceive his theory in a geometrical way, something that came only later with his embrace of the explicitly geometrical notion of parallel transport or what we know today as the affine connection. According to Guttfreund and Renn, for instance, commenting on these sections in their annotated edition of the 1916 paper, Einstein in 1916 neither "systematically introduce[d] non-Euclidean geometry, nor did he interpret his own theory in terms of differential geometry" (Gutfreund and Renn 2015, 81). Similarly, according to Ryckman

> the concept of parallel transference of a tangent vector in a Riemannian manifold *M* was first developed in 1917 by Levi-Civita ... The hitherto purely analytical Christoffel symbols (of the second kind) of covariant differentiation are equated with the components (relative to a given coordinate system) of the unique affine (henceforth "Levi-Civita") connection associated with the metric. This furnishes the Christoffel symbols with a geometrical interpretation, by relating them to the parallel displacement of a vector along a path ....[17]

But if Einstein in 1916 treats the Christoffel symbols and covariant differentiation in a purely analytical way, does it follow that his subsequent embrace of parallel transport marks a turn to a more geometrical conception?

I am doubtful on this point. Parallel transport is, to be sure, a geometrical concept, but the concept of a line element (*ds*) is equally geometrical. Moreover, it is not clear that Einstein's subsequent embrace of parallel transport was motivated by geometrical considerations. In the Princeton lectures he introduces covariant differentiation with the comment that such operations are "recognized most satisfactorily in the ... way, introduced by Levi-Civita and Weyl" (Einstein 1922, 69), but he does not explain why parallel transport is to be preferred to the method employed in the 1916 paper. Perhaps Einstein merely regarded the earlier procedure as more cumbersome. Einstein's remarks on the matter in 1955 are more enlightening:

> It is well known that around the turn of the century Riemann's theory of metrical continua, which had fallen so completely into oblivion, was revivified and deepened by Ricci and Levi-Civita; and the work of these two decisively advanced the formulation of general relativity. However it seems to me that Levi-Civita's most important contribution lies in the following theoretical discovery: the most essential theoretical accomplishment of general relativity, namely, the elimination of "rigid" space, i.e. of the inertial system, is only indirectly connected with the introduction of a Riemannian metric. The immediately essential conceptual element is the "displacement field" (Γ^l_{ik}) which expresses the infinitesimal displacement of vectors.[18]

Clearly Einstein's concern in the passage above is not with geometry but with the elimination of the inertial frame. Indeed, his reasoning appears equivalent to that in the letter to Laue from five years earlier (quoted above Sect. 7.2.1.1) on the rigidly rotating disk: what characterizes the existence of a gravitational field is the non-vanishing of the Γ^l_{ik}, not the non-vanishing of R_{iklm}. That is, the Riemannian metric is incapable of distinguishing between a "flat" spacetime for which the Riemann tensor vanishes and a genuine gravitational field relative to an accelerated frame in a finite Galilean region (for which the Riemann tensor also vanishes). Only the affine connection, derivable independently of the Riemannian metric and vanishing solely for a flat spacetime, can register the distinction. Thus if anything this line of argumentation, which highlights the centrality of the principle of equivalence to the general theory of relativity, points *away* from a geometrical interpretation rather than toward one.

Whatever the evolution of Einstein's thinking on the issue of the geodesic law of motion, whether a metric geodesic based on stationary ds or its equivalent in terms of parallel transport of the four-velocity vector, from our perspective the crucial point is that either method is on its face geometrical: stationary ds because it employs the notion of a line element and parallel transport because it employs the notion of parallelism. We shall see below, however, that stationary ds can be conceived in a more physically intelligible way apart from the geometrical notion of a geodesic line. Parallel transport, by contrast, is an essentially geometrical notion.

7.2.2.3 Part B, §12: The Riemann Tensor

We consider next Einstein's analysis of the Riemann tensor in Part B, section 12. With respect to our reading of the paper thus far the obvious question is why, unless that theory is geometrical in a substantive rather than merely formal sense, the mathematical machinery of Riemannian geometry should be so effective. The answer is that the relativistic expression $g_{\mu\nu} dx_\mu dx_\nu$ is exactly analogous in a formal sense to the line element of differential geometry, the two differing solely by the negative sign of the spatial variables in the relativistic version. But this difference is formally irrelevant for purposes of the requisite mathematical manipulations, since we are interested not in the values of the variables x_n per se but rather solely in the $g_{\mu\nu}$ and its derivatives with respect to those variables.

In the sense of the general theory of relativity, Einstein highlights in §12 the specific significance of the Riemann tensor, constructed from $g_{\mu\nu}$ and its first and second derivatives:

> The mathematical importance of this tensor is as follows: If the continuum is of such a nature that there is a coordinate system with reference to which the $g_{\mu\nu}$ are constants, then all the $B^{\rho}_{\mu\sigma\tau}$ [Riemann tensor] vanish. If we

choose any new system of coordinates in place of the original ones, the $g_{\mu\nu}$ referred thereto will not be constants, but in consequence of its tensor nature, the transformed components of $B^{\rho}_{\mu\sigma\tau}$ will still vanish in the new system. Thus the vanishing of the Riemann tensor is a necessary condition that, by the appropriate choice of the system of reference, the $g_{\mu\nu}$ may be constants. In our problem this corresponds to the case in which, with a suitable choice of the system of reference, the special theory of relativity holds good for a *finite* region of the continuum. (141)

What do we learn from this passage? If Einstein were speaking of a geometrical theory defined in terms of a line element we would learn that when the Riemann tensor vanishes we have a Euclidean or "flat" space. But in the context of relativistic physics we learn something quite different: If the Riemann tensor vanishes, Einstein notes, then by an appropriate choice of coordinates the special theory of relativity holds in a finite region. That appropriate choice of coordinates is in fact the choice of an inertial frame of reference coordinatized with rigid rods and clocks, such that the geometry of space is Euclidean and clocks at different locations can be synchronized. But, if for the same condition—the vanishing of the Riemann tensor in a finite special relativistic region—we choose an accelerated system of coordinates (the rigidly rotating disk, for example), then instead we obtain a gravitational field of a special kind. In terms of this field, (1) the geometry of *space* will be non-Euclidean and (2) we will no longer be able to synchronize clocks at different locations such that the difference in time coordinates has a direct metrical significance. What we do *not* learn from the passage is that the Riemann tensor informs us of the "geometry of spacetime." We have gotten so used to metaphorical ways of speaking that we forget what the theory is actually telling us.

If we wish to speak of the "Riemann tensor" in the context of general relativity we should above all be cognizant of the fact that in so speaking we refer to a purely analytical entity, albeit one originally invented in the context of differential geometry. But if it is understood as a geometrical entity then the "Riemann tensor" is not the $B^{\rho}_{\mu\sigma\tau}$ of Einstein's theory, as articulated in the passage quoted above. For in the very text before us (Part B, section 12), Einstein in fact obtains $B^{\rho}_{\mu\sigma\tau}$ "from the fundamental tensor alone" and the fundamental tensor ($g_{\mu\nu}$) is neither derived from differential geometry nor in any substantive way related to Minkowski's theory.

7.2.3 *Einstein 1916, Part C: Theory of the Gravitational Field*

We focus in this part on the derivation of the law of motion in §13, the vacuum field equation in §14, and the general field equation in §16.

7.2.3.1 Part C, §13: Law of Motion

The mathematical apparatus for the equation of motion already having been given in terms of stationary $\int ds$ in section 9 of Part B, Einstein now proceeds to the physical application of the mathematical concept of a geodesic. With respect to Einstein's derivation of the law of motion in Part C, section 13, we must underline once again that the use of "*ds*" to represent the left-hand member of $\sqrt{g'_{\sigma\tau}dx'_{\sigma}dx'_{\tau}} = \sqrt{g_{\sigma\tau}dx_{\sigma}dx_{\tau}}$ is unjustified in a physical sense, it not having been shown that this expression has anything to do with a line element. Here I propose an equivalent derivation using extremal proper time rather than extremal *ds*, following Einstein's account in section 13 and inserting the appropriate alterations as we proceed.

Einstein begins by recalling the law of inertial motion for finite special relativistic regions and the possible extension of that law to the general theory of relativity:

> A freely movable body not subjected to external forces moves, according to the special theory of relativity, in a straight line and uniformly. This is also the case, according to the general theory of relativity, for a part of the four-dimensional space in which the system of coordinates K_0 may be, and is, so chosen that they have the special constant values given in (4). (Einstein 1952b, 142)

By "they" in the final clause Einstein evidently means the metric coefficients $g_{\sigma\tau}$ specified in Part A, §4, where he noted that for an appropriately chosen coordinate system in a finite special relativistic region those coefficients take on the constant diagonal values −1, −1, −1, +1. Let us return temporarily to the equation derived in that section, which should properly be written not as $ds^2 = g_{\sigma\tau}dx_{\sigma}dx_{\tau}$, but rather $g'_{\sigma\tau}dx'_{\sigma}dx'_{\tau} = g_{\sigma\tau}dx_{\sigma}dx_{\tau}$. Writing it out longhand, with the *y* and *z* dimensions suppressed, yields

$$g'_{11}c^2dt'^2 + g'_{44}dx'^2 + 2g'_{14}ct'x' = g_{11}c^2dt^2 + g_{44}dx^2 + 2g_{14}ctx$$

We can then assign the constant metric coefficients −1 and 1 as appropriate to obtain the special relativistic equation $-x'^2 + c^2t'^2 = -x^2 + c^2t^2$, which comes directly from Einstein 1905 special relativity. We demonstrated in Chap. 6 that this equation is in reality a symbolic abbreviation for a proportion, and so the two algebraic quantities equated ($-x'^2 + c^2t'^2$ and $-x^2 + c^2t^2$) are themselves nominalized ratios rather than physical or geometrical quantities in their own right.

Keeping the preceding observation in mind, let us stipulate an inertial state of motion for the primed frame such that the two events in question occur at the same point in space. We may, therefore, rewrite the equation as $c^2\tau^2 = -x^2 + c^2t^2$, where τ designates the time recorded on a clock at rest in the primed frame (thus registering coordinate time in the primed frame). Let us suppose, in addition, that this clock is present at both of the events in question. Clearly, τ^2 is proportional to $-x^2 + c^2t^2$, and so by substituting for the consequents of our ratios the unit of time and the unit of distance respectively, we obtain the proportion

$$\tau : unit\,\tau :: \sqrt{\left(-x^2 + c^2t^2\right)} : unit\,x.$$

Since our clock is free of external forces and initially at rest relative to the primed frame, the law of inertia dictates that it remain at rest relative to the primed frame. Thus, were the clock to experience any acceleration during the time interval between the two events, such that it departed and subsequently returned to its original location, it would register a special relativistic time dilation effect and the elapsed time recorded by the clock along its trajectory would be less than the elapsed time had it remained at rest.

Hence, Einstein's 1905 special relativity, in concert with the law of inertia, implies that *proper time is maximized for an inertial trajectory.* Moreover, so must the symbolic quantity $\sqrt{-x^2 + c^2t^2}$ measured in some other inertial frame be maximized, since this quantity is proportional to proper time. Clearly the proportional holds irrespectively of whether $\sqrt{-x^2 + c^2t^2}$ is itself a physical or geometrical quantity, which it is not. Thus we now have a law of free body motion for the special theory of relativity: $\delta\int_A^B\left(-dx^2 + c^2dt^2\right) = 0$. No conception of a "line element" has entered into our derivation; for all we are interested in is the fact that the symbolic quantity $\sqrt{-x^2 + c^2t^2}$ is maximized, because it is *proportional to proper time.* That is, it is not maximized because it is a "straight line" in four-dimensional spacetime or because it represents a spacetime distance. Consequently, the special relativistic law of motion should be referred to stationary proper time, and not to stationary *ds* as in Einstein's account. The two are not the same, even if they are proportional. Stationary *ds* is in fact a physically meaningless concept, since an inertial trajectory is a straight line solely on a graph. Happily, we can dismiss all that and hold fast to the physically intelligible concept of maximal proper time recorded on a clock. Such a clock does not "measure the interval," as is often said, but rather *measures its frame-time* and, in so doing, also measures indirectly the proportional algebraic quantity $-x^2 + c^2t^2$.

Having determined the law of motion in special relativity by referring it to stationary proper time instead of stationary *ds*, let us pick up again with Einstein's account in section 13 of Part C. Still assuming a finite Galilean region, Einstein observes that on the basis of the principle of equivalence we can refer this motion in "any chosen system of coordinates" to a gravitational field of a special type. Relying on the geodesic equation introduced in Part B, §9, Einstein therefore writes, for this special gravitational trajectory,

$$\frac{d^2x_\sigma}{d\tau^2} = \Gamma^\sigma_{\mu\nu}\frac{dx_\mu}{d\tau}\frac{dx_\nu}{d\tau}$$

There is no Minkowski spacetime geometry here, the Christoffel symbols being purely analytical entities to derive the equation of motion on the basis of maximal proper time. In fact, given that the equation of motion is intended to represent the acceleration of a falling body, or the rate of change of its velocity over time, it is best to write the equation in terms of coordinate time, dispensing altogether with the four-velocity vector. In physical terms, after all, velocities and accelerations are referred to coordinate time rather than to a proper time parameter. Hence, following Clifford Will, our law of motion is expressed more revealingly in terms of coordinate time as follows:[19]

$$\frac{d^2x_\sigma}{dt} = \Gamma^\sigma_{\mu\nu}\frac{dx_\mu}{dt}\frac{dx_\nu}{dt} - \Gamma^4_{\mu\nu}\frac{dx_\mu}{dt}\frac{dx_\nu}{dt}\frac{dx_\sigma}{dt}.$$

Thus, as we noted in Chap. 2 for the special relativistic case, all relevant information contained in a Minkowski four-vector is determined by the three-dimensional spatial components. In the present case the gravitational trajectory is completely determined by the spatial part $\frac{d^2x_j}{dt^2}$, for the time component reduces to[20]

$$\frac{d^2t}{dt^2} + \Gamma^4_{\alpha\beta}\frac{dx_\alpha}{dt}\frac{dx_\beta}{dt} - \Gamma^4_{\alpha\beta}\frac{dx_\alpha}{dt}\frac{dx_\beta}{dt}\frac{dt}{dt} = 0 .$$

Since both sides of this equation vanish, regardless of the gravitational potentials contained in the Christoffel symbols, the time component of the four-dimensional law of motion contains no information about the gravitational trajectory. Moreover, the original $\frac{d^2t}{d\tau^2} + \Gamma^\mu_{\alpha\beta}\frac{dx_\alpha}{d\tau}\frac{dx_\beta}{d\tau} = 0$, with proper time parameter, is derivable from the coordinate time version above, and so the proper time version in terms of the four-velocity contains no additional information about the trajectory. We conclude that the geodesic equation is in substance really three-dimensional and that the time component $\frac{d^2t}{d\tau^2}$ is not a vector component in any physically meaningful sense. We have thus freed ourselves of the physically obscure notion of a static gravitational trajectory in four-dimensional spacetime, replacing it with the more physically intelligible notion of a trajectory unfolding through time in three-dimensional space.

Einstein next makes the transition to gravitational fields associated with source masses, where no finite Galilean region is available:

> We now make the assumption, which readily suggests itself, that this covariant system of equations also defines the motion of the point in the gravitational field in the case when there is no system of reference K_0, with respect to which the special theory of relativity holds good in a finite region. (143)

The assumption to which Einstein refers, however, is truly reasonable only if we hold fast to the principle of equivalence in Einstein's sense, for otherwise we would have no reason to believe that an equation of motion for a fictitious gravitational field in a finite special relativistic region would apply to a "real" gravitational field associated with source masses, nor could we extrapolate based on what transpires in an infinitesimal region in which special relativity holds.[21]

We have arrived at the law of motion for the general theory of relativity without recourse to Minkowski spacetime. We therefore shall have discharged the task of this chapter if we similarly can show that the gravitational field law itself no more relies on Minkowski's theory than does the law of motion.

7.2.3.2 Part C, §14: Vacuum Field Law

Einstein commences his derivation of the vacuum field law by once again invoking the principle of equivalence and the special gravitational fields associated with it:

> Here we again apply the method employed in the preceding paragraph in formulating the equations of motion of the material point. A special case in which the required equations must in any case be satisfied is that of the special theory of relativity, in which the $g_{\mu\nu}$ have certain constant values. Let this be the case in a certain finite space in relation to a definite system of coordinates K_0. Relatively to this system all the components of the Riemann tensor $B^{\rho}_{\mu\sigma\tau}$ ... vanish. For the space under consideration they then vanish, also in any other system of coordinates. (143–144)

Our desired gravitational field law, then, must at a minimum accommodate the special gravitational fields induced by accelerated reference frames in finite special relativistic regions, since these are real gravitational fields. Otherwise we would simply have learned how to formulate special relativity in generally covariant fashion, but with no reason to require the law of gravitation to cover such special gravitational fields. Were we to limit ourselves to just those special fields our gravitational field law would be $B^{\rho}_{\mu\sigma\tau} = 0$. But the gravitational field law must cover also those fields, associated with matter sources, which cannot be "transformed away" by a change of coordinates. Therefore, while the vanishing of the Riemann tensor is a sufficient condition for the satisfaction of the law of gravity, it is not a necessary condition:

> Thus the required equations of the matter-free gravitational field must in any case be satisfied if all the $B^{\rho}_{\mu\sigma\tau}$ vanish. But this condition goes too far. For it is clear that, e.g., the gravitational field generated by a material point in its environment certainly cannot be "transformed away" by any choice of the system of coordinates, i.e., it cannot be transformed to the case of constant $g_{\mu\nu}$. (144)

As is well-known, Einstein's choice for a gravitational tensor other than the Riemann tensor is strictly constrained at this point by the requirement that the desired tensor be constructed out of $g_{\mu\nu}$ and its derivatives no higher than second order. He fixes upon the only feasible candidate, the

Ricci tensor $R_{\mu\nu}$ (Einstein here calls it $G_{\mu\nu}$), formed by contraction of the Riemann tensor.

This choice yields the vacuum field law $R_{\mu\nu} = 0$, which Einstein announces with evident satisfaction:

> It must be pointed out that there is only a minimum of arbitrariness in the choice of these equations, for besides $R_{\mu\nu}$ there is no other tensor of second rank which is formed from the $g_{\mu\nu}$ and its derivatives, contains no derivatives higher than second, and is linear in these derivatives. (144)[22]

Since Einstein reiterates that in his derivation of the field law he has "proceeded by the method of pure mathematics," (144–145) it is difficult to see any reliance on Minkowski's theory, since the mathematical apparatus comes rather from Riemannian geometry. The only possible connection to Minkowski would be the metric tensor $g_{\mu\nu}$, on the assumption it entails the concept of a line element; but, as we have already observed, this entity is simply a generalization of a set of coefficients (−1, −1, −1, +1) already present in the special relativistic equation $-dx_1'^2 - dx_2'^2 - dx_3'^2 + dx_4'^2 = -dx_1^2 - dx_2^2 - dx_3^2 + dx_4^2$. Clearly, the strictly analytical procedure of rendering the preceding equation generally covariant cannot in itself make it geometrical or in any way dependent upon Minkowski's theory.

Nevertheless, it is understandable that the employment of a metric tensor would be mistaken for an embrace of Minkowski spacetime. For the concept of a metric tensor originates in differential geometry and in that context is defined essentially in terms of a quadratic distance function. It is then natural to assume, and indeed might even seem to be self-evident, that if in the general theory of relativity we employ a metric tensor $g_{\mu\nu}$, formally analogous to the metric tensor of differential geometry, it too must be defined in terms of a distance function, but in this case a "spacetime" distance rather than a spatial distance per se as in differential geometry itself. But this line of reasoning is simply wrong. The distance function of differential geometry is derived from a geometrical theorem (the Pythagorean Theorem) which has to do with relations between geometrical squares built on sides of a right triangle. The analogous relativistic expression $-dx_1^2 - dx_2^2 - dx_3^2 + dx_4^2$ is derived rather from the Lorentz transformation and could be a "generalization" of the Pythagorean Theorem only if the Lorentz transformation itself were a generalization of the Pythagorean Theorem. But as we concluded last chapter, it is not.

The doubt is sure to persist, nevertheless, that if in Einstein's theory the metrical properties of space and time are represented ***in a single equation*** ($c^2 d\tau^2 = g_{\sigma\tau} dx_\sigma dx_\tau$) those metrical properties surely have been merged into a single continuum as envisioned by Minkowski's theory. But once again the conclusion simply does not follow. Already in Einstein's 1905 special relativity the metrical properties of space and time are jointly represented, by means of a single equation, without recourse to the notion of a space-time distance. Metrical properties are all those properties that can be measured, and in the special theory of relativity what the existence of a single equation jointly expressing the metrical properties of space and time tells us is that those metrical properties are entangled, not that they are merged into a single continuum. Only in geometry proper are all dimensions merged into a single continuum.

7.2.3.3 Part C, §16: General Field Equation and Stress-Energy Tensor

For the completion of our analysis there remains the role of the stress-energy tensor in Einstein's general field equation. Here the theory of Minkowski spacetime would lead us to believe that we require for the source of the gravitational field not merely a relativistic expression for mass-energy, as we might have expected, but rather some generalization of the momentum four-vector of Minkowski's theory. This ends up forcing upon us additional components, such as pressure and stress, which we might not have thought should contribute to the gravitational field. We shall presently see that an energy tensor cannot be derived from the Minkowski four-momentum and that, in any event, $T_{\mu\nu}$ does not represent an "invariant geometrical object." Indeed, we might do well to dispense with $T_{\mu\nu}$ altogether. Einstein's own dissatisfaction with the role of the energy tensor in his theory of gravity is well-documented. In his essay "Physics and Reality" of 1936, Einstein compares his field equation to "a building, one wing of which is made of fine marble (left part of the equation), but the other wing of which is built of low-grade wood (right side of equation [$T_{\mu\nu}$])."[23] I believe, however, the situation for $T_{\mu\nu}$ is even worse than Einstein's bleak assessment would suggest: $T_{\mu\nu}$ is comparable not to low-grade wood, to press the metaphor, but rather to a wing with no foundation at all.

Textbook accounts of general relativity, which as a rule avoid mention of Einstein's own doubts about the energy tensor, typically inform us that for the right-hand side of the Einstein equation we need to include as sources for the gravitational field *all* forms of energy, as if for some reason we had up to now been ignoring some of those forms. We soon find ourselves in possession of a tensor $T_{\mu\nu}$ that includes among its components not just energy, but also pressure, stress, momentum density, and so forth, which in fact are not forms of energy at all. As a rule, the usual account of the stress-energy tensor unfolds with no explanation of why such non-energy components should act as sources of gravity in the first place.

In fact, the structure of $T_{\mu\nu}$ is dictated not by the need to include "all sources of energy" but by two quite different, albeit related, considerations: First, since mass-energy is not invariant in special relativity it seems as if the invariant four-momentum, the time component of which is interpreted as energy, must be our point of departure for an energy tensor. Otherwise one might have expected, on the analogy with Newton's theory of gravity, that the source of curvature should be energy density alone. Secondly, since the left-hand side of the Einstein equation consists of a two-index (sixteen component) tensor $G_{\mu\nu}$, it evidently stands to reason that the right-hand side must also consist of a second rank tensor. This inference evidently is buttressed by the fact that in the special theory of relativity we do require a second rank tensor to represent energy density, with two frame-dependent corrections required, one for relativistic mass and one for a relativistic volume contraction.

In accord with established practice, then, let us adopt the Minkowski four-momentum as our point of departure for an energy tensor and focus initially on the time component $m\frac{d(ct)}{d\tau}$ or $mc\gamma$. Things go awry right from the start. The time component of the Minkowski four-momentum, as we observed in Chap. 2, reduces not to E (relativistic energy) but to $\frac{E}{c}$, no physical quantity in its own right at all, energy or otherwise. For once again, we cannot simply make the c in the denominator "go away" by setting the velocity of light to unity. Hence, if the T^{44} component of the stress-energy tensor truly is to represent energy density, it cannot be derived from the Minkowski four-momentum.

Our textbook accounts as a rule first introduce $\rho u^t u^t$ as the energy density component of the stress-energy tensor for dust (where ρ is rest mass density and u^t is the time component of Minkowski four-velocity). Of course, since u^t is $\frac{d(ct)}{d\tau}$ or $c\gamma$, what we really have obtained is simply $\rho c^2\gamma^2$ or the relativistic energy density. And since that is the very quantity we would have desired for the energy density in the first place, one wonders why the detour through the four-velocity, other than to assure ourselves that we are proceeding in truly "four-dimensional" terms. In any case, the relativistic energy density $\rho c^2\gamma^2$ can in no wise be derived from the time component of the Minkowski four-momentum, for in that event the T^{44} component instead would be $\rho c\gamma^2$, which is not energy density. From a physical perspective, then, Minkowski spacetime is hereafter out of the picture except notationally.

With a formulation in hand for energy density, we next learn in our standard accounts that this very quantity is best regarded as the flux of energy through a "hypersurface of constant time" or, more generally, the flux of the μth component of the four-momentum across a surface of constant ν, yielding again for dust $T^{\mu\nu} = \rho u^\mu u^\nu$. The initial and wrong-headed commitment to the Minkowski four-momentum now compels us to define the other fifteen components of the stress-energy tensor in terms of such "fluxes." Accordingly, the $T^{4\nu}$ components yield energy fluxes across surfaces of constant x, y, or z, and the $T^{\mu 4}$ components give rise to "momentum densities" or fluxes of momentum in the time direction. Unfortunately for the theory, if these six aforementioned components are to carry the same units as energy density, which they must if they are to be set equal to the components of the Einstein tensor on the left-hand side of the Einstein equation, then an arbitrary divisor of c must be inserted into energy flux and an arbitrary factor of c into momentum density. For clearly a density of energy ($\frac{mc^2}{volume}$) cannot have the same units as either a density of momentum ($\frac{mv}{volume}$) or a flux of energy ($\frac{mc^2}{area - time}$). We are usually told that momentum density is "the same as" energy flux because it has the same units, a singularly unhelpful explanation of the former entity.

Let us proceed to the remaining nine "inner" components of $T^{\mu\nu}$, which comprise the "stress tensor." Once again, in terms of fluxes, we introduce the flux of x, y, or z momentum in the x, y, and z directions respectively. Thus for the components T^{xx}, T^{yy}, and T^{zz} we have pressures and for the remaining cross-product terms shear stresses. But difficulties once again immediately assert themselves. In a perfect fluid (zero viscosity and heat conductance) the shear stresses drop out and we are left with pressure alone. But pressure, which *ex hypothesi* characterizes the fluid element under consideration, is not a flow of momentum, whatever that would mean. Rather, the so-called "flow" of momentum is a purely mathematical entity or symbolic quantity with the same units as pressure.

Furthermore, the physically accurate assertion that a perfect fluid, although it exhibits no shear stresses, still exerts pressure stems not from the concept of "fluxes" of the components of the Minkowski four-momentum, but rather is derived directly from physical considerations relating to the motion of particles. Once again, when we encounter a physically intelligible component of the stress-energy tensor we find that it is unrelated to four-vectors and when we find a connection to four-vectors the component thus obtained is physically unintelligible.

The stress-energy tensor $T_{\mu\nu}$ conceived as a generalization of the Minkowski four-momentum is beset with all the conceptual and mathematical anomalies discussed above. These very anomalies raise the further question, which has been looming over our entire discussion, of why pressure, stress, or any of the other components besides energy density should be regarded as sources of gravity in the first place. That is to say, what other reason besides the presence of a second-rank tensor on the left-hand side of Einstein's field equation—admittedly no mean consideration—would cause us to think that any tensor components besides mass-energy should make a contribution to the gravitational field? Even for the standard cosmological case of a perfect fluid, for instance, there is no experimental evidence of any such additional contribution to gravity by pressure. Based on purely physical considerations, geometrical objects aside, if pressure, for instance, is to act as a source of gravity independently of energy density, then it should also so act in vacuum solutions to Einstein's field equation. Why then does the Schwarzschild exterior metric contain no term for pressure?[24] With respect to the issue of such a separate pressure contribution to gravity, Vishwakarma acutely remarks that

> the term ρ [mass-energy density] includes in it all the possible sources of mass and energy (excluding gravitational energy). Hence, so is included in it the energy equivalent to the work done against external pressure. If we already know that energy (in the form of work done) is being supplied to the system or getting released from it ... why can't this too be taken care of by the term ρ? There is no natural law which dictates that ρ cannot include particular types of energies.[25]

Vishwakarma concludes, "It is thus established that the relativistic description of matter [for a perfect fluid] ... suffers from some subtle inherent inconsistencies in its basic formulation."[26] That is from our perspective rather an understatement.

What needs to be questioned is the very assumption that we require an energy tensor on the right-hand side of Einstein's gravitational field equation. The presence of non-energy terms like pressure in the stress-energy tensor suggests the possibility that the right-hand side of Einstein gravitational field equation should not be occupied by any tensor at all. Here Einstein's demonstration of the Newtonian weak-field limit for the non-vacuum field equation is especially revealing. In the 1916 review article, Part E, §21, Einstein observes that for the weak field limit we need consider, in the equation of motion $\frac{d^2x_\tau}{ds^2} = \Gamma^\tau_{\mu\nu}\frac{dx_\mu}{ds}\frac{dx_\nu}{ds}$, only those terms for which $\mu = \nu = 4$ (where $t = x_4$); for we deal in this case with velocities small in comparison to the velocity of light. Therefore, the equation of motion, which assumes $s = t$ for small velocities, reduces to $\frac{d^2x_\tau}{dt^2} = \Gamma^\tau_{\mu\nu}$.[27] Moreover, on the assumption that for a quasi-static field differentiations with respect to the time coordinate are small compared with differentiations with respect to the space coordinates, $\Gamma^\tau_{\mu\nu}$ reduces to $-\frac{1}{2}g^{\tau\alpha}\left(\frac{\partial g_{44}}{\partial x_\alpha}\right)$.[28] Further reduction with $\alpha = \tau$ and $g^{\tau\tau} \cong 1$ (since we have no off-diagonal metric coefficients and the diagonal ones differ only slightly from unity) yields $\frac{d^2x_\tau}{dt^2} = \frac{1}{2}\frac{\partial g_{44}}{\partial x_\tau}$ ($\tau = 1, 2, 3$ since differentiation with respect to time vanishes). Therefore, without recourse to the field equation we have derived the Newtonian equation of motion, with $\frac{1}{2}g_{44}$ fulfilling the role of the gravitational potential in Poisson's equation. The Newtonian limit is determined to a first approximation by g_{44} alone, on the basis of the general relativistic geodesic law.

Einstein now reverses his procedure and recovers the field equation for the Newtonian limit from the weak-field metric above. The left side of the field equation reduces to $G_{44} = -\frac{1}{2}\left(\frac{\partial^2 g_{44}}{x_1^2} + \frac{\partial^2 g_{44}}{x_2^2} + \frac{\partial^2 g_{44}}{x_3^2}\right)$ or $-\frac{1}{2}\nabla^2 g_{44}$. Moreover, with energy density alone to consider in light of our assumptions, the energy tensor reduces to $T_{44} = \rho$. Hence, we appear to have obtained the Newtonian limit $\nabla^2 g_{44} = k\rho$ on the basis of Einstein's field equation $G_{\mu\nu} = kT_{\mu\nu}$. But not really. Rather, what Einstein has done is simply substitute the left-hand side $G_{\mu\nu}$ of his field equation into the left-hand side of the Poisson equation $\nabla^2\Phi = k\rho$. It is quite arbitrary to label ρ as T_{44} and, in fact, the energy tensor has played no role in the derivation whatsoever. It is rather the erroneous construction of the stress-energy tensor on the basis of Minkowski's four-momentum, which equates energy with the time component of the four-momentum, that gives the appearance of a physical rationale for pairing G_{44} with T_{44}, when in reality none exists. The Newtonian limit Einstein has actually derived, therefore, is the vacuum equation $\nabla^2 g_{44} = 0$, not the more general $\nabla^2 g_{44} = k\rho$. Then, on the basis of $\nabla^2 g_{44} = 0$, we can deduce Newton's inverse-square law and, going backwards from the inverse-square law, finally obtain the general version of the Newtonian limit $\nabla^2 g_{44} = k\rho$. The stress-energy tensor, therefore, does not come into play except notationally in Einstein's derivation of the Newtonian limit of his field equation.

But that is not all. From the beginning the only reason we sought out a general equation in the form $G_{\mu\nu} = kT_{\mu\nu}$, instead of being satisfied with the "pure" field equation $R_{\mu\nu} = 0$ ($R_{\mu\nu}$ designating the Ricci tensor), is that (1) the analogy with Newtonian field law $\nabla^2\phi = k\rho$ suggested we should seek for a non-vacuum analogue in general relativity, and (2) the energy density ρ of the Newtonian theory therefore appeared to require reformulation as a matter tensor. But if the substance of Einstein's recovery of the Newtonian limit is simply the reduction of $G_{\mu\nu}$ to G_{44}, which relies solely on the left-hand side of the field equation, why regard $R_{\mu\nu} = 0$ as a "vacuum field law" in the sense of a matter-free case of a more general field equation? For in the present case, once we have determined that *any* Newtonian weak field reduction, vacuum or non-vacuum, is governed by $G_{44} = -\frac{1}{2}\left(\frac{\partial^2 g_{44}}{x_1^{\ 2}} + \frac{\partial^2 g_{44}}{x_2^{\ 2}} + \frac{\partial^2 g_{44}}{x_3^{\ 2}}\right)$, we know also that our gravitational law for weak fields reduces to an inverse-square relation. On that basis we solve for the non-vacuum weak field ($\nabla^2 g_{44} = k\rho$) with no stress-energy tensor. The preceding variation clearly is not sufficient for non-vacuum solutions in general, but it raises the possibility that the so-called "vacuum field equation" (or what Einstein came to prefer as the "pure gravitational field") might be *the* field equation rather than merely a matter-free special case of a more general field equation.

7.2.3.4 The Gravitational Field Equation Without a Matter Tensor

To make the proposal above more plausible, consider Einstein's remark, at the conclusion of Section §4 of his 1916 article, on the significance of zero tensors in the formulation of generally covariant laws of physics:

> The things hereafter called tensors are further characterized by the fact that the equations of transformation are linear and homogeneous. *Accordingly, all the components of the new system vanish, if they all vanish in the original system. If, therefore, a law of nature is expressed by equating all the components of a tensor to zero, it is generally covariant* [my italics].[29]

Einstein's emphasis on zero tensors might seem misplaced, since any law of nature can be written in such a way that it equates all the components of a tensor to zero. In the specific case of the Einstein equation we can write instead of $G_{\mu\nu} = kT_{\mu\nu}$ the equivalent $G_{\mu\nu} - kT_{\mu\nu} = 0$, which equates all of the components of the tensor $G_{\mu\nu} - kT_{\mu\nu}$ to zero. What then is the particular significance of zero tensors as far as general covariance is concerned?

To appreciate Einstein's line of thinking in §4 we need to recall that Einstein regards the requirement of general covariance as a kind of "mathematical sieve," by means of which the true field law can be discovered by eliminating all, or at least almost all, competing possibilities. Certainly Einstein was correct in assigning to general covariance this heuristic role, in the sense that the Ricci tensor is in fact the only tensor obtainable from the metric tensor and its derivatives and of no higher than second differential order. But Einstein's particular logic of discovery can work here only if we are looking for a zero tensor from the beginning. That is, although $G_{\mu\nu} - kT_{\mu\nu}$ is indeed a zero tensor, Einstein could never have found it directly based on the requirement of general covariance, since there are innumerable other tensors obtainable in this fashion. Rather, we must from the start seek for a tensor that meets the mathematical criterion of Einstein's "sieve." Then we can set all the components of this tensor to zero for a generally covariant field law of gravitation. And only after the Ricci tensor has been obtained in this way can we introduce a matter tensor on the right-hand side of the field equation and modify the left-hand side in accordance with energy conservation (that is, zero divergence of the stress-energy tensor) to yield $R_{\mu\nu} - \frac{1}{2} g_{\mu\nu} R = kT_{\mu\nu}$.

Consequently, Einstein's search for a zero tensor should not be understood as a search for a "vacuum field law" per se, but rather in the first place as a search for general covariance, in accordance with the heuristic role Einstein assigns the latter in the formulation of a theory of gravity. Or, as Einstein puts it in his Autobiographical Notes, we should focus on the law of the "pure gravitational field," with respect to which the matter source formulation on the right-hand side is "merely a makeshift in order to give the general principle of relativity a preliminary closed-form expression."[30] On the possibility we are suggesting, the pure gravitational field equation $R_{\mu\nu} = 0$ would be *the* field equation itself, with non-vacuum solutions consigned to the status of special cases.

7.3 Geodesic Law by Other Means

We now return to the question of parallel transport because it may be fairly asked how this unmistakably geometrical notion, in the present case clearly associated with the Minkowski four-velocity vector, should "work," so to speak, in the general theory of relativity. But the question can be disposed quickly, since we answered it in essence already in Chap. 3 in the context of special relativity, where we observed that a "straight line in four-dimensional spacetime" is a graph of the equation for an inertial trajectory. Since the law of inertia dictates $\delta\left(\frac{dx_\mu}{dt}\right) = 0$, our law of geodesic motion must be $\frac{d^2x_\mu}{dt^2} = 0$, and so, in terms of geometrical representation on a spacetime graph, if we think of a four-dimensional "velocity vector" tangent to the graph of an inertial trajectory, then obviously that vector must always lie on the graphed straight line and so be parallel to itself. The situation is in principle no different for a "most inertial" trajectory in the general theory of relativity. Here the law for "parallel transport" of the four-dimensional velocity vector is $\delta\frac{(dx_\mu)}{d\tau} = -\Gamma^\mu_{\alpha\beta}\frac{dx_\alpha}{d\tau}dx_\beta$, such that $\frac{d^2x_\mu}{d\tau^2} + \Gamma^\mu_{\alpha\beta}\frac{dx_\alpha}{d\tau}\frac{dx_\beta}{d\tau} = 0$. The representation is geometrical only in the sense that parallel transport is a geometrical concept; the Christoffel symbols themselves have a purely analytical sense. Einstein therefore expresses himself unfortunately, in his Princeton lectures, when he suggests that in the theory of gravity geodesic motion is "the simplest generalization of the straight line."[31]

From a somewhat similar perspective on Minkowski's theory as ours, Harvey Brown is keen to emphasize that the law of motion in general relativity can be derived directly from the field equation (even from the vacuum field equation in Einstein's preferred "singularity" method).[32] By this means, on Brown's line of thought, no explanatory role need be assigned Minkowski spacetime in the derivation of the general relativistic law of motion. From our perspective, of course, Brown's worries are ill-founded, since we have argued that neither stationary $d\tau$ nor parallel transport ultimately relies on the concept of Minkowski spacetime. I remain skeptical, though, that a direct derivation of the geodesic law of motion from the field equation has the significance Einstein attached to it, in the sense of obviating the need for an independent hypothesis of motion in the general theory of relativity. In the version of the direct derivation cited by Brown, for instance, the law of motion is obtained based on the vanishing of the covariant divergence of the stress-energy tensor for the freely falling test body. However, the vanishing of the covariant divergence of Einstein's field equation is the result of applying the law of conservation of energy and momentum to the stress-energy tensor on the right-hand side. That is, we add the trace term $-\frac{1}{2}Rg_{\mu\nu}$ on the left in order to satisfy conservation on the right. But if we must apply conservation of energy to derive the field equation in the first place, why make the detour through the field equation instead of deriving the law of inertia directly from the principle of conservation of energy? And why should we regard an independent law of inertia more distastefully than we would an independent law of conservation of energy? From our perspective, then, it is hard to see a great advantage in deriving the geodesic law directly from the field equation.

Notes

1. Norton 1993, 799–800.
2. Page numbers otherwise unidentified in this section refer to Einstein's 1916 paper "The Foundation of the General Theory of Relativity" (Einstein 1952b [1916]).
3. The definitive discussion of the question is Norton 1989.
4. Einstein 1952b [1916], 143.
5. Einstein to Laue, September 12, 1950. Quoted in Norton 1989, 39–40.
6. Synge 1960, ix.
7. Stachel 1989a, 58.
8. Einstein 1961 [1916], 109–110.

9. Einstein 1961 [1916], 105. On the point-coincidence argument see Stachel 1989b.
10. In a letter to Lorentz of January 1916 (Einstein 1998, 232–233), Einstein mentions three reasons for the downfall of the *Entwurf* theory: (1) it did not accommodate rotating frames of reference, (2) incorrect perihelion motion of Mercury, (3) the maximal degree of covariance he was able to achieve failed to uniquely determine the field equation.
11. According to Norton (1987, 155), for instance, it is a "commonplace of differential geometry ... [that] any well formulated spacetime theory is automatically expressible in coordinate free (= generally covariant) terms."
12. Gutfreund and Renn 2015, 183. Unfortunately, the introduction to the 1916 review article is omitted from the Perrett and Jeffery English translation. An updated English translation which includes the introduction is provided in Gutfreund and Renn, 183–232.
13. Poincaré had already noticed it at least three years prior to Minkowski's 1908 paper "Space and Time," but Einstein does not appear to have been aware of Poincaré's precedence in this respect.
14. Einstein 1952b [1916], 119.
15. Einstein 1953 [1922], 11.
16. Einstein 1953 [1922], 12.
17. Ryckman 2005, 150.
18. Quoted in Gutfreund and Renn 2015, 81.
19. Will 1993, chapter 6.
20. For a metric geodesic in special relativity, the time component $\frac{d^2t}{d\tau^2}=0$ informs us merely that the relativistic factor γ is a constant, something that in any event follows necessarily from the space terms. If we then substitute $\frac{dt}{\gamma}=d\tau$, we obtain $\frac{dx^\alpha}{dt^2}=0$. Clearly the time term $\frac{d^2t}{dt^2}=0$ is uninformative, with all the information regarding the trajectory of a free body contained in the three space terms.
21. With regard to the "infinitesimal principle of equivalence," there is no way of articulating a law of geodesic motion at all for an infinitesimal region of space and time since all trajectories, including non-geodesic ones, approach geodesic "straightness" in the infinitesimal. As Einstein writes to Moritz Schlick, based on the straightness of a trajectory in the infinitesimal "nothing can be derived.... [For] in the infinitesimal every continuous line is straight" (*Heiraus kann aber nichts abgeleitet werden.... im Unendlichkleinen ist jede stetige Linie eine Gerade*) (Einstein to Schlick, March 21, 1917; Einstein 1998, 417). Or in the more picturesque formulation of Torretti, to argue for a geodesic law of motion based on infinitesimal straightness "is like arguing that the parallels of latitude on the surface of the Earth are no less geodesic than the meridians, because they agree, for instance,

with the avenues that cross Chicago from East to West, which are no less straight [in the infinitesimal] than those leading from North to South" (Torretti 1983, 316). Norton, 1989 (sections 9–10, 31–39), includes an informed discussion of this point.

22. It is sometimes maintained that the restriction to a second rank tensor derives from the stress-energy tensor, which is of second rank and to which the gravitational tensor must be equated in the general field law. However, the restriction to second rank can be proved mathematically (Einstein 1922, 84) and does not depend on characteristics of the stress-energy tensor.
23. Einstein 1982 [1954], 311. Similarly, in his "Autobiographical Reflections" of 1949 Einstein remarks, "The right-hand side is a formal condensation of all things whose comprehension in the sense of a field theory is still problematic. Not for a moment, of course, did I doubt that this formulation was merely a makeshift in order to give the general principle of relativity a preliminary closed-form expression" (Einstein 1979 [1949], 71).
24. Arnowitt et al. (2008 [1962], 2024) treat the absence of pressure as a source term in the interior Schwarzschild solution with the remark that pressure is registered rather as "clothing of the original mechanical mass." This suggests that pressure is subsumed by the energy density component of the stress-energy tensor and should not appear as a separate component in its own right.
25. Vishwakarma 2012, 376.
26. Vishwakarma 2012, 377.
27. With $c = 1$, the four-displacement s is equal to the proper time τ, and so for small velocities $s = t$.
28. $\Gamma^{\tau}_{44} = \frac{1}{2} g^{\tau\alpha} \left(\frac{\partial g_{4\alpha}}{\partial t} + \frac{\partial g_{4\alpha}}{\partial t} - \frac{\partial g_{44}}{\partial x_{\alpha}} \right).$
29. Einstein 1952b [1916], 121.
30. Einstein 1979 [1949], 71.
31. Einstein 1953 [1922], 79.
32. Brown 2005, 161–163. Also Brown and Pooley 2006, sec I.

PART IV

Time Without Spacetime

CHAPTER 8

Relativity and Time

Many philosophers regard temporal becoming as irreconcilable with both special and general relativity.[1] In special relativity, it would seem, distant simultaneity is either conventional, frame-relative, or both. Therefore, coming to be, entailing as it does a present in which things become, must render indeterminate the coming to be of distant events: becoming either will be limited to a local "now" or relativized to frames of reference with differing simultaneity relations. Furthermore, a given solution to Einstein's gravitational field equation presents indefinitely many possible sets of simultaneous events or "foliations" of time, evidently once again rendering the coming to be of distant events indeterminate. Static theories of time, on the other hand, which reject tensed descriptions of physical reality, appear to receive a kind of scientific *imprimatur* from Minkowski's theory. I conclude this study, then, with some observations on the status of time and becoming in a theory of relativity without spacetime.

8.1 Simultaneity in Special Relativity

In his analysis of simultaneity in the 1905 paper on special relativity, Einstein treats local simultaneity as self-evident, except for a footnote: "We shall not here discuss the inexactitude which lurks in the concept of two events at approximately the same place, which can only be removed by an abstraction."[2] Two events occurring at *approximately* the same place

J. K. Cosgrove, *Relativity without Spacetime*,
https://doi.org/10.1007/978-3-319-72631-1_8

in principle would still require a clock synchronization protocol. Presumably, then, the abstraction Einstein has in mind is the mathematical idealization by which we regard two events as exactly coinciding at a spatial point. That idealization is surely justified in the context of Einstein's analysis, but since the occurrence of two events at a single mathematical point is a physical impossibility, in the philosophy of time we cannot accept local simultaneity so defined as a primitive or self-evident concept. Indeed, we might well regard distant simultaneity in the vicinity as the more empirically transparent notion. After all, Einstein formulates his definition of local simultaneity in experiential terms—a direct perception of simultaneity in my "immediate proximity." What I actually experience, however, is not simultaneity in the idealized sense (two events occurring at a single point in time and a single point in space), but rather what is sometimes called the "specious present" or the time corresponding to the minimum threshold of human consciousness, extending spatially to all the events I can experience within that duration.

Local simultaneity is often regarded as a *relativistic* invariant, but this is a misleading characterization since simultaneity is a topological relation and topological relations are frame-independent.[3] Local simultaneity is not a relativistic invariant like the speed of light, for instance, which is inherently frame-relative but happens to be the same in all frames. Should there be such a thing as distant simultaneity, we would expect it to be frame-independent as well. Unfortunately, Einstein somewhat confuses the question in his 1905 paper by first declaring distant simultaneity conventional, only to subsequently classify it as frame-relative. Thus Einstein's frame-relative simultaneity is itself conventional. As Winnie shows, for instance, we can eliminate disagreement on simultaneity between frames in relative motion by selecting alternate clock synchronization conventions.[4] In any event, there can be no question that Einstein meant to deny the reality of distant simultaneity, not merely render it frame-relative. As late as 1949, for example, Einstein affirms, as a "definitive result" of the special theory of relativity, that "[t]here is no such thing as simultaneity of distant events."[5]

While the debate on conventionality continues in the literature, only an unjustified operationalism would equate our inability to *measure* the one-way transit time of light with the conventionality of that one-way transit time and, along with it, the conventionality of distant simultaneity itself. However, Winnie and more recently Robert Rynasiewicz, for instance, have argued that the conventionality of distant simultaneity is non-trivial in special relativity, and that, notwithstanding Einstein's mode of expression in 1905, the thesis does not have its origin in an operationalistic concept of science. According to Winnie, the arbitrariness of distant clock synchronization within a single frame

> reveals a structural feature of the Special Theory, and thereby of the universe it purports to characterize, which not only makes the one-way speed of light indeterminate, but reveals that its unique determination could only be at the expense of *contradicting* the non-conventional content of the Special Theory .... Thus the CS [conventionality of simultaneity] thesis should not be seen as having its basis in an operationalistic view of the nature of physical theory, but rather as pointing to a structural feature of the Special Theory which this theory does not share with classical kinematics.[6]

An example of the non-conventional content Winnie is referring to would be round-trip clock retardation. While one-way clock retardation is empirically unverifiable and depends on the arbitrary choice of a clock synchronization protocol (and in fact can be eliminated through choice of the appropriate protocol), round-trip retardation is empirically verifiable and independent of synchrony conventions. Winnie concludes that this "structural feature" of the special theory of relativity, the independence of non-conventional content with respect to synchrony conventions, would be inconsistent with the determination of a unique or "true" synchrony.

It is hard to see just where Winnie sees a conflict between the synchrony-independence of special relativistic effects such as round-trip clock retardation and the operationalistic reading of the conventionality thesis. For suppose that in reality the one-way speed of light were equal to its average round-trip speed. Non-standard synchrony would still yield the same round-trip clock retardation as standard synchrony, since the differences in the one-way transit times exactly compensate for each other. And if what Winnie specifically means to proscribe is merely any unique *empirical* determination of the one-way transit time of light, at least by methods employing finite speeds (slow clock transport, for instance), we can assent to that proposition without foreclosing the possibility of a true one-way speed of light.

Rynasiewicz offers a somewhat more provocative account of the conventionality thesis in terms of diffeomorphism equivalence.[7] His proposal essentially amounts to designating events with the same time coordinate as simultaneous, and then reassigning events to different manifold points, such that previously simultaneous events are no longer simultaneous. That is, we began with standard clock synchrony as per Einstein 1905, and obtain non-standard synchrony in the diffeomorphically related model (and, in turn, end up with standard synchrony in the new coordinate system corresponding to the diffeomorphism). According to Rynasiewicz, by analogy with the standard resolution of the hole argument in general relativity, we are to regard any two simultaneity conventions related by a diffeomorphism as physically equivalent. This conclusion seems premature, though, for we must first determine the extent to which the analogy holds between the hole argument and the problem of distant simultaneity. At issue in the hole argument is our freedom to distribute the metric field over the manifold in many different ways. The argument for physical equivalence in this case is that the alternative is a singularly strange type of physical indeterminism that both fails to express itself in anything observable and involves the dubious existence of bare (abstract) manifold points. In the case of distant simultaneity, however, there is no similarly compelling argument for physical equivalence. To be sure, a unique simultaneity relation would be empirically unverifiable within the constraints of the theory of relativity. However, that is just what we should expect given the status of light as a "first signal." Moreover, it is unclear what is even to be understood by the assertion that light has a determinate "average" round-trip speed, but no determinate instantaneous speed. For what is being averaged if not the instantaneous speed of light at each point of its round-trip? Unlike the hole scenario, then, nothing could be more natural than for light to have a determinate one-way speed equal to its round-trip speed.

What we can say, in sum, is that distant simultaneity cannot register in special relativity. In this methodological sense, there is no distant simultaneity in special relativity and we thus are compelled to regard distance simultaneity as conventional within the round-trip transit time of light. That is, once again, a methodological constraint, not a settled judgment as to what is the case in nature. Moreover, if simultaneity is a strictly topological relation it should not be frame-relative. *Relative* simultaneity at a distance is not part of the empirical core of special relativity, and we have every reason to reject the concept. Finally, special relativity is valid only infinitesimally in general relativity, so relative simultaneity at a distance is no longer even a consideration when we consider gravity. Thus, in view of its methodological constraints, we can only conclude that the special theory of relativity leaves open the question distant simultaneity.

8.2 Simultaneity in General Relativity

In special relativity, we encountered two potential sources of arbitrariness in the determination of distant simultaneity: (1) the conventionality of simultaneity in a single frame of reference, predicated on the conventionality of one-way light velocity, and (2) the relativity of simultaneity among inertial frames of reference in relative motion. We must ask whether either of these impediments is applicable to general relativity.

With respect to the latter, no general principle of relativity is realized in Einstein's theory of gravity, except in the hypothetical scenario underwritten by the principle equivalence in a finite special relativistic region. Einstein was well aware that the general principle of relativity did not license us to "transform away" the gravitational field of the sun, for instance, in a finite region.[8] Thus, as Kretschmann showed early on in the history of the general theory, we must carefully distinguish between general coordinate freedom (general covariance) and a general relativity principle underwriting use of arbitrary frames of reference. In the special theory of relativity, by comparison, Lorentz coordinate transformations or "boosts" correspond to a change of inertial reference frame. Therefore, even though special relativity may be formulated in generally covariant terms, the theory must be *at least* Lorentz covariant. But the most we can say in general relativity is that its field equation must accommodate, at a minimum, the degree of covariance associated with the principle of equivalence—for instance, rigidly rotating frames in a finite special relativistic regions (one of the downfalls of the *Entwurf* theory).

Since the general principle of relativity, underwritten by the principle of equivalence, does not obtain in our actual world, Einstein's theory of gravity realizes no principle of relativity such as could underwrite the relativity of simultaneity among reference frames in relative motion, at least not in the presence of gravitational fields associated with source masses. The idea that it does reflects a failure to distinguish properly between coordinate systems and reference frames. Merely by changing coordinates we do not adopt a new reference frame in the sense entailed by a relativity principle; for given a reference body in a particular state of motion we can always erase the old coordinates, so to speak, and inscribe new ones, thereby changing coordinate systems without changing reference frames. By contrast, the arbitrary "foliations" of time for a solution to Einstein's field equation are all defined within a single coordinate system. To be sure, the differential rate of clocks in a gravitational field precludes the employment the clock synchronization protocol of special relativity, where coordinate differences correspond directly to time intervals. But in some cases, at least, we can correct for the differential rate of clocks, so it would be wrong-headed to insist that simultaneity is strictly local in general relativity. The global positioning system (GPS) adjusts the rate of satellite clocks so they synchronize with clocks on the surface of the earth, such that truly simultaneous ticks of the respective clocks have the same time coordinate.[9] We could always be stiff-necked and insist that since the GPS synchronization is dependent on a one-way light velocity assumption, the clock ticks are simultaneous in a conventional sense only. But this species of conventionality, as we saw, can be regarded in terms of diffeomorphism; and, as we argued above, there is no strong case for diffeomorphic equivalence in the case of distant simultaneity. Rather, we encounter once again, now in general relativity, a methodological constraint on our determination of the one-way speed of light.

What we do learn from general relativity is that gravity has a metrical effect on clocks and that a free clock measures the time interval between any two events at which it is present. In Newton's theory of absolute time, if we start with a set of globally simultaneous events then a given time interval determines another set of globally simultaneous events. The same also holds in special relativity given a simultaneity convention within a single frame. Thus we could say that in both Newton's theory and in special relativity, we can determine topological time (the simultaneity relation) as a function of a global metrical time: in Newton's theory absolute time "flows equably" and in special relativity we assume that identically constructed clocks run at the same rate when at rest. This connection between topological and metrical time does not obtain in general relativity, since the rate at which processes occur (the ticking of a clock, for instance) depends on the gravitational field. Nevertheless, global simultaneity is consistent with general relativity.

Since simultaneity is a topological feature of time, and topological properties in general are frame-independent, it seems wrong-headed to ground the concept of global simultaneity in a preferred cosmological reference frame, as is sometimes attempted through considerations of the homogeneity and isotropy of the universe. For one thing, all we would thereby secure would be an average; and, just as there exists in the real world no "average American," neither should there exist in the physical universe an average "now." What cosmology can furnish, rather, is an approximation of the *non*-frame-relative present on a global scale, such that we can, for instance, speak of the present age of the universe. If clocks did not on average tick at the same rate, so to speak, everywhere in the universe, the universe as a whole would not have an age. Similarly, the cosmic microwave background at present has the same temperature everywhere because it all started in the same place and has been cooling for the same amount of time. But all of this *presupposes* global simultaneity. We do not need a preferred cosmological reference frame to underwrite the concept of global simultaneity. In general relativity, the "preferred frame" is the local gravitational field itself, relative to which metrical effects such as time dilation and the Lorentz contraction may be referred. Such a preferred frame does not presuppose cosmological homogeneity and isotropy.

With respect to the question of whether or in what sense time can be regarded as "absolute," Michael Friedman usefully distinguishes three senses of the term absolute in the debate on the nature of space and time: (1) absolute versus *relational*, in the sense of the Leibniz-Newton debate over the independent reality of space and time; (2) absolute versus *frame-relative* (best regarded as a particular form of the preceding distinction, since a frame-relative property is a relation to a frame); and (3) absolute versus *dynamical*, in the sense of Einstein's stricture on so-called absolute objects that act but are not acted upon.[10] With regard to the first of Friedman's contrasts, we must embrace a relational account according to which time is essentially bound up with the passage of events in the physical world. Clearly, in the theory of relativity we cannot endorse a definition of "absolute time" such as we find in Newton's famous scholium, where time "flows equably" in itself, independently of the passage of events. With respect to frame-relativity, our analysis suggests that while metrical time is frame relative, topological time is global and frame-independent. On Friedman's third contrast, between absolute and dynamical objects, we affirm the dynamical character of metrical time. However, metrical time is a matter of the comparative rates at which different processes occur, as when we compare the rate of some process to the rate of a clock. Comparative rates do not directly affect the order of events. A clock ticks more slowly in a gravitational field, but the positions of its hands are successively ordered the same way they would have been without gravity. In this sense, then, topological time is not dynamical. Yet the differential rates at which processes occur obviously affects which particular distant events are simultaneous with which other ones, and in this sense topological time is dynamical as well. Topological relations in time seem more fundamental than metrical relations, for a metric interval presupposes two events already ordered earlier and later in time. This very precedence of topological time is presumably why we can subjectively "lose track of time" in the metrical sense, but not (at least normally) in the topological sense. If I am immersed in something I cease to compare the rate of my activity to the rate of the clock on the wall, for instance, and so I am surprised when what seemed like twenty-minutes was actually three hours. But I do not usually find myself surprised that an event I thought occurred beforehand actually occurred afterward.

8.3 Time and Becoming

The problem of distant simultaneity in the theory of relativity has suggested to some authors the possibility of a theory of local becoming. Dennis Dieks, for instance, proposes a conception of tensed time based on the "successive coming into being of events" at a particular location.[11] Becoming in Dieks's sense is local in that (1) successive events come to be along a world-line, and (2) temporal ordering relations are non-global. However, such a theory of local becoming, which has much else to recommend it, evidently saves becoming at the cost of what Vesselin Petkov has aptly called "event solipsism."[12] For if becoming is defined in terms of events successively *coming into being* locally, then the being of events comprising distant processes of becoming is rendered indeterminate. Moreover, as we already noted regarding Einstein's 1905 discussion of local simultaneity, the concept of the local present is ambiguous in itself, since "here and now" in the strict sense is a mathematical abstraction. R.T.W. Arthur argues for an extension of the local present ("specious present"), the time of mutual interaction, to a round trip of light within the minimum duration of a conscious experience. One virtue of such a view is its comportment with our experiential intuitions—you and I can actually have a cup of coffee together in a shared "now." But in the end, the problem of the reality of distant processes of becoming is not satisfactorily disposed by any theory of the specious present, since we have merely replaced "event solipsism" with what we might call "regional solipsism." On the theory of the specious present, that is, events outside the present's spatial extent are still indeterminate as to their being.

Along the lines of Dieks and Arthur, though, let us conceive an event as the *act* of some entity—a being or substance in the traditional terminology. We do not conceive of an event atomistically, but think rather of an event arising out of and perhaps overlapping with an immediately preceding event, and extending into an immediately succeeding event. The concept of a point-event being a mathematical abstraction, clearly in the concrete an event must have some finite temporal duration. The question cannot be avoided, therefore, as to what *unifies* the duration of an individual event or makes all its temporal "moments" part of that one duration. On this question, Leibniz argued that both time and space are unreal because, as infinitely divisible mathematical continua, they lack substantial unity. He instead posited a "formal atom" (monad) or acting substance underlying the phenomenal world in time and space.[13] We need not accept Leibniz' metaphysical schema to acknowledge the force of his argument regarding unity: duration must be more than a mere succession of temporal points, but mere succession of points is the only aspect of the relation "before and after" that registers in the theory of relativity.

Where does the present time or "now" enter our picture? We observed earlier that the experienced "now," while it does not register in the theory of relativity per se, is presupposed in Einstein's very definition of local simultaneity. We can with advantage avail ourselves of Aristotle's definition of change (*kinēsis*) or becoming in terms of the actualizing of a potential for being: becoming, that is, in Aristotle's famous definition from *Physics*, is the "actualization of the potential qua potential."[14] That is to say, before the event has occurred the potential is not yet *being actualized*, and after the event has occurred it has already been actualized. Therefore, an event or action occurs solely in the present, the time when that which was merely potential is *being* actualized but has not yet been actualized. Only when the event or action *has become*, according to Aristotle, can we say it is actual and therefore no longer itself still a process:

> At the same time we see and have seen, understand and have understood, think and have thought; but we cannot at the same time learn and have learned, or become healthy and be healthy. We are living well and have lived well, we are happy and have been happy, at the same time; otherwise the process would have had to cease at some time … but it has not ceased …[15]

Only in the *perfect tense*, as it were, is an event determinate or actualized: for then it already is and so *has become*.[16]

Insofar as the theory of relativity depends on the concept of "events," then from a philosophical point of view we have every reason to hold to the reality of becoming. Static theories of time are unable to give any coherent account of what an event actually is; and it would be extraordinary if an event itself were an "illusion," as static theories of time inevitably imply. However, becoming does entail simultaneity at a distance—unless we are willing to countenance local event solipsism—and simultaneity at a distance simply cannot register in relativistic physics per se. Craig Callender aptly sums up the situation:

> Tensers are wasting their time trying to find an image of the tensed theory in physics. Specific theories will be more or less hostile to tenses, but in general they will be against tenses so long as there is no clear need for them. Show physics a need for tenses and it will quickly accommodate them. Until then, merely as a by-product of scientific methodology, physics will not accommodate them.[17]

I believe Callender is quite right. However, there is no reason to believe that the tensed time of experience it is illusory simply because it does not register in mathematical physics. Furthermore, we need to be clear on the precise sense in which the failure of tensed time to register in physics is a "by-product of scientific methodology."

The modern concept of science is coined above all in the writings of Descartes, especially his early treatise on method, *Rules for the Direction of the Mind*. In Rule 4 of that work, Descartes specifies the subject matter of his projected "universal mathematics" (*mathesis universalis*) as "order and measure":

> When I considered the matter more closely, I came to see that the exclusive concern of mathematics is with questions of order or measure [*ordo vel mensura*] and that it is irrelevant whether the measure in question involves numbers, shapes, stars, sounds, or any other object whatever. This made me realize that there must be a general science which explains all the points that can be raised concerning order and measure irrespective of the subject matter, and that this science should be termed *mathesis universalis*—a venerable term with a well-established meaning—for it covers everything that entitles these other sciences to be called branches of mathematics.[18]

The "other sciences" to which Descartes refers, which qualify as branches of mathematics only because they too are concerned exclusively with order and measure, are the traditional "mixed sciences," or what Aristotle in *Physics* II.2 calls the "physical parts of mathematics": optics, music, astronomy, and so forth. Whatever are the other attributes or particular "matter" of the objects of mathematics, mathematics itself examines strictly order and measure to the exclusion of all else. Descartes' point is strictly methodological here: a "certain science," as he calls his projected enterprise, must deal exclusively with order and measure.[19] In later and more familiar texts such as the *Discourse on Method* and *Meditations*, Descartes adds significant metaphysical apparatus to the methodological principle of the *Rules*: now the very being of the physical world is purely "extension," for instance, and only mathematical properties such as size, shape, motion, and the like are truly real. However, the later Cartesian metaphysics of extension is in large part polemical, an assault on the regnant Aristotelian science with its substantial forms, final causes, and so forth, sterile notions all of them, which Descartes wishes to purge from his new science.

The efficacy of the Cartesian scientific methodology, which is the methodology of modern science in general, can hardly be doubted now some four centuries into the modern scientific era. But the metaphysical apparatus is not essential to the scientific methodology. For comparison's sake, consider Descartes' elimination from biology of the concept of "soul" as principle of life. Descartes in fact goes so far as to argue for the elimination of any distinction between a sick body and a healthy body in the science of biology:

> Those who are ill, for example, may desire food or drink that will shortly afterwards turn out to be bad for them. Perhaps it may be said that they go wrong because their nature is disordered, but this does not remove the difficulty. A sick man is no less one of God's creatures than a healthy one, and it seems no less a contradiction to suppose that he has received from God a nature which deceives him. Yet a clock constructed with wheels and weights observes the laws of nature just as closely when it is badly made and tells the wrong time as when it completely fulfills the wishes of the clockmaker. In the same way, I might consider the body of man as a kind of machine ...[20]

While modern biology has fully embraced the Cartesian clock metaphor, the science of medicine surely cannot dispense with an intrinsic distinction between a sick body and a healthy body, even if for methodological reasons that distinction does not register in biology per se. Rather, the distinction is presupposed in biology, just as the experienced "now" is presupposed in the relativistic definition of local simultaneity. Likewise, mathematical physics cannot itself deny the reality of tensed time, even if in view of its methodological commitments it cannot register that reality. In fact, physics takes for granted the reality of becoming every time it speaks of a physical *process*. As Aristotle observed at the beginning of Western science, no special science demonstrates its own principles.

What, ultimately, is the *time* of mathematical physics but *order* (topological time) and *measure* (metrical time)? Methodologically, then, tensed time does not and cannot register in the science of physics. In this sense, Einstein was quite right to assert that "[t]he concept [of simultaneity] does not exist for the physicist until he has the possibility of discovering whether or not it is fulfilled in an actual case."[21] Such utterances have sometimes been attributed to philosophical positivism on Einstein's part, at least in this stage of his career, but they are better regarded as methodological.

Since the methodologically reduced time of special and general relativity leaves open the question of distant simultaneity, we should affirm global and non-frame-relative simultaneity as the most philosophically coherent position. The "eternalist" theory, or theory of static time, tries to have it both ways philosophically by affirming the reality of time in the sense of serial order while regarding time as unreal in its essential character as becoming. But McTaggart rightly pointed out that we cannot affirm the *B* series (non-tensed ordering of events) without also affirming the *A* series (becoming of events); for time essentially involves change, and becoming is the only form temporal change could assume. Without becoming we cannot ultimately distinguish the temporal ordering of events from their spatial ordering. Certainly a plus sign rather than a minus sign in the metric signature is not enough to accomplish that. Any distinguishing mark of temporal change (directionality, causality, and the like) implies the reality of becoming.

In rejecting the static view of time, however, we do not necessarily embrace so-called presentism, at least not in an unqualified sense.[22] The present obviously does enjoy a privileged reality, as when we say of someone who has died that "he is no more" or of a loved one who had been in pain, "Thank goodness she is no longer suffering." In another sense, though, both the future and the past are real: the future *as* future and the past *as* past. Probably we do not get very far by disputing about whether the present is more real than the past and future.[23] The crucial question is rather the reality of becoming, which we have every reason to affirm.

Notes

1. I shall generally avoid the term "tensed time," which is often interpreted in the philosophy of time as synonymous with becoming. In his informative overview of time in special relativity, Savitt (2011, 563) distinguishes between tensed time and becoming, suggesting that the latter can be understood strictly in terms of "serially ordered clock times." If by becoming we mean coming into being, though, as the word suggests, then becoming inevitably involves tense.

2. Einstein 1952a [1905], 39.
3. In this connection, one of the difficulties with Hilary Putnam's famous argument that all events, whether past, present, or future, are "equally real," since one can always designate a frame of reference relative to which a given event is simultaneous with some event occurring "now," is that Putnam casts local simultaneity as a frame-relative invariant rather than a topological absolute. While it is true, if we must so put it, that two locally simultaneous events are simultaneous "in all frames," since local simultaneity is in fact a frame-independent invariant, it is not true that locally simultaneous events occur "at the same time" in all frames, in the sense of occurring at a time common to frames in relative motion. Rather, a moment of time in a given frame is determined by the plane of simultaneity relative to that frame, and two frames with different planes of simultaneity have no moments in common; nor does a point at which the respective planes intersect define a *common moment*. For at this point in space we simply have two simultaneous events, without any relation to reference frames. The "I-now/You now simultaneity-at-a-point" adduced by Putnam in support of the transitivity of the "as real as" relation for events not lying on a single plane of simultaneity (Putnam 1967, 242) is in fact a function of the insertion of a conscious observer who perceives a "now" at the point in question. But this "now" is the subjective now of the observer, which in the context of the special theory of relativity does not determine a temporal moment common to frames in relative motion. In the sense of the special theory of relativity, after all, an observer's subjective "now" would be simply injected into the manifold as one more event alongside all the others, in no wise bestowing a common "now" upon inertial frames in relative motion. Moreover, in special relativity inertial reference frames are not defined by "observers" at rest in them in the first place, and in fact observers per se do not register in special relativity at all. Therefore, no "I-now/You now" at a point can underwrite the transitivity of Putnam's "as real as" relation.
4. Winnie 1970, 84.
5. Einstein 1979 [1949], 57.
6. Winnie 1970, 81–82.
7. Rynasiewicz 2012.
8. See for example Part C, §14 of Einstein's 1916 review article (Einstein 1952b [1916], 144): "Thus the required equations of the matter-free gravitational field must in any case be satisfied if all the $B^{\rho}_{\mu\sigma\tau}$ vanish. But this condition goes too far. For it is clear that, e.g., the gravitational field generated by a material point in its environment certainly cannot be 'transformed away' by any choice of the system of coordinates, i.e. it cannot be transformed to the case of constant $g_{\mu\nu}$."

9. On GPS, see Tom Van Flandern's (2008) analysis of GPS and relativistic effects.
10. Friedman 1983, 62–64.
11. Dieks 2006b, 172.
12. Petkov 2005, 124.
13. Leibniz' argument on substantial unity is in various texts, such as "A New System of Nature" (1695): "Now, a multitude can derive its reality only from *true unities*, which have some other origin and are considerably different from points which all agree cannot make up the *continuum*. Therefore, in order to find these *real entities*, I was forced to have recourse to a formal atom, since a material thing cannot be both material and, at the same time, perfectly indivisible, that is, endowed with true unity" (Leibniz 1989 [1695], 139).
14. For instance *Physics* III.1, 201a10.
15. Aristotle, *Metaphysics* IX.6, 1048b25–30; Aristotle 1953, 449.
16. Scholars of Aristotle will note that the present passage properly refers to the fulfillment of an activity or a perfect actualization (*entelecheia*). However, for Aristotle the account of change in terms of perfect actualization is the most intelligible that can be given and so is paradigmatic for change in general, even if most forms of change fall short of perfect actualization.
17. Callender 2008, 67.
18. Descartes 1985–1991 [ca. 1628], 1:19.
19. It is not entirely clear in the *Rules* how Descartes means to distinguish between "order" and "measure." In *Principles of Philosophy*, a later and in many ways more mature work, Descartes associates order with *number*, stressing that "we should not regard order or number as anything separate from the things which are ordered or numbered" (Descartes 1985–1991 [1644], I:211). In the same passage Descartes also mentions duration in time, which would obviously fall under "measure." Order evidently has to do with how things are arranged, including numerical arrangement such as the succession of moments of time in correlation with what we now call real numbers.
20. Descartes 1985–1991 [1641], 1:58.
21. Einstein 1961 [1916], 26.
22. See the revealing discussion on this point by Unger (Unger and Smolin 2015, 245–248 and 518–521), although I do not agree that if "the now has no unique value in natural science" (521) then the now has no privileged status in natural philosophy. It is for methodological reasons, in my judgment, that the now has no privileged status in physics itself.
23. A nice discussion of the question of the reality of the present versus the reality of past and future is Dorato 2006.

Erratum to: Relativity without Spacetime

Joseph K. Cosgrove

Erratum to:

The Author(s) 2018
J. K. Cosgrove, *Relativity without Spacetime*,
https://doi.org/10.1007/978-3-319-72631-1

The original version of this book was revised. The corrections are updated in this revised version.

The updated original online version for this book can be found at https://doi.org/10.1007/978-3-319-72631-1

J. K. Cosgrove, *Relativity without Spacetime*,
https://doi.org/10.1007/978-3-319-72631-1_9

Bibliography

Aristotle. 1953. *Metaphysics I–IX*. Translated by H. Tredennick. Cambridge, MA: Harvard University Press.

Arnowitt, Richard, Stanley Deser, and Charles W. Misner. 2008 [1962]. Republication of: The Dynamics of General Relativity. *General Relativity and Gravitation* 40 (9): 1997–2027. https://doi.org/10.1007/s10714-008-0661-1.

Arthur, R.T.W. 2006. Minkowski Spacetime and the Dimensions of the Present. In Dieks 2006a, 129–155.

———. 2008. Time Lapse and the Degeneracy of Time: Gödel, Proper Time and Becoming in Relativity Theory. In Dieks 2008, 207–227.

Balashov, Yuri, and Michel Janssen. 2003. Critical Notice: Presentism and Relativity. *British Journal for Philosophy of Science* 54: 327–346. https://doi.org/10.1093/bjps/54.2.327.

Barbour, Julian. 1999. *The End of Time: The Next Revolution in Physics*. Oxford: Oxford University Press.

Bashmakova, I.B. 1977 [1972]. *Diophantus and Diophantine Equations*. Updated by Joseph Silverman. Translated from the Russian by Abe Shenitzer with the editorial assistance of Hardy Grant.

Bell, J.S. 1987. How to Teach Special Relativity. In *Speakable and Unspeakable in Quantum Mechanics*, edited by J.S. Bell, 67–80. Cambridge: Cambridge University Press.

Bohm, David. 1996 [1965]. *The Special Theory of Relativity*. London: Routledge.

Bondi, Hermann. 1980 [1964]. *Relativity and Common Sense: A New Approach to Einstein*. New York: Dover. Republication of original edition, New York: Doubleday & Company, Inc.

J. K. Cosgrove, *Relativity without Spacetime*,
https://doi.org/10.1007/978-3-319-72631-1

Brown, Harvey R. 2005. *Physical Relativity: Space-time Structure from a Dynamical Perspective*. Oxford: Oxford University Press.

Brown, Harvey R., and Oliver Pooley. 2006. Minkowski Space-Time: A Glorious Non-Entity. In Dieks 2006a, 67–89.

Callender, Craig. 2008. Finding 'Real' Time in Quantum Mechanics. In Craig and Smith 2008, 50–72.

Corry, Leo. 1997. Hermann Minkowski and the Postulate of Relativity. *Archive for History of Exact Sciences* 51 (4): 273–314. http://www.jstor.org/stable/41134033.

———. 1998. The Influence of David Hilbert and Hermann Minkowski on Einstein's Views Over the Interrelation Between Physics and Mathematics. *Endeavor* 22 (3): 95–97. https://doi.org/10.1016/S0160-9327(98)01131-4.

Cosgrove, Joseph. 2012. On the Mathematical Representation of Spacetime: A Case Study in Historical-Phenomenological Desedimentation. *New Yearbook for Phenomenology and Phenomenological Philosophy* XI: 154–186.

Craig, William Lane, and Quentin Smith, eds. 2008. *Einstein, Relativity and Absolute Simultaneity*. Oxford: Routledge.

Damour, Thibault. 2008. What Is Missing from Minkowski's 'Raum und Zeit' Lecture. *Annalen der physic* 17 (9–10): 619–630. https://doi.org/10.1002/andp.200810315.

Descartes, René. 1954 [1637]. *The Geometry of Descartes, with a Facsimile of the First Edition*. Translated by Eugene Smith and Marcia L. Latham. New York: Dover. An unabridged and unaltered reproduction of the English translation originally published by Open Court Publishing Co. in 1925.

———. 1985–1991. *The Philosophical Works of Descartes*. Translated by John Cottingham, Robert Stoothoff, Dugald Murdoch, and Anthony Kenny. 3 vols. Cambridge: Cambridge University Press.

———. 2001 [1637]. *Optics, Geometry and Meteorology*. Translated with Introduction by Paul J. Olscamp. Rev. ed. Indianapolis: Hackett.

Dieks, Dennis, ed. 2006a. *The Ontology of Spacetime*. Amsterdam: Elsevier.

———. 2006b. Becoming, Relativity, and Locality. In Dieks 2006a, 157–176.

———, ed. 2008. *The Ontology of Spacetime II*. Amsterdam: Elsevier.

Diophantus. 1893. *Opera Omnia*. Editit et Latine interpretatus est Paulus Tannery. 2 vols. Leipzig: B.G. Teubner.

———. 2009. *Arithmetic*. Translated by Thomas L. Heath. In Heath, 129–259.

Dorato, Mauro. 2006. The Irrelevance of the Presentist/Eternalist Debate for the Ontology of Minkowski Spacetime. In Dieks 2006a, 93–109.

During, Elie. 2012. On the Intrinsically Ambiguous Nature of Space-Time Diagrams. *Spontaneous Generations* 6 (1): 160–171. https://doi.org/10.4245/sponge.v6i1.18645.

Eddington, Arthur. 1965 [1923]. *The Mathematical Theory of Relativity*. London: Cambridge.

Einstein, Albert. 1952a [1905]. On the Electrodynamics of Moving Bodies. In Einstein et al 1952, 35–65.

———. 1952b [1916]. The Foundation of the General Theory of Relativity. In Einstein et al 1952, 109–164.

———. 1953 [1922]. *The Meaning of Relativity*. 5th ed. Princeton: Princeton University.

———. 1961 [1916]. *Relativity: The Special and General Theory*. 15th ed. Translated by Robert W. Lawson. New York: Wings Books.

———. 1979 [1949]. *Autobiographical Notes*. Edited and translated by Paul Arthur Schilpp. La Salle: Open Court Press. First published in Schilpp 1949.

———. 1982 [1954]. *Ideas and Opinions*. Based on *My Weltbild*, edited by Carl Seelig, and other sources. New translations by Sonja Bargmann. New York: Three Rivers Press.

———. 1998. *The Collected Papers of Albert Einstein*, vol. 8, Part A, edited by Robert Schulmann, A.J. Cox, Michel Janssen, and József Illy. Princeton: Princeton University.

———. 2002 [1920]. Fundamental Ideas and Methods of the Theory of Relativity, Presented in Their Development. In *The Collected Papers of Albert Einstein*, vol. 7 (English Translation Supplement), edited by Michel Janssen, Robert Schulmann, József Illy, Christof Lehner, and Diana Kormos Buchwald, translated by Alfred Engel, 113–150. Princeton: Princeton University Press. http://einsteinpapers.press.princeton.edu/vol7-trans/. Accessed 15 Aug 2017.

Einstein, Albert, H.A. Lorentz, Hermann Weyl, and Hermann Minkowski. 1952. *The Principle of Relativity*. Translated by W. Perrett and G.B. Jeffery, with notes by A. Sommerfeld. New York: Dover.

Euclid, T.L. 1956. *The Elements*. Translated with Introduction by Sir Thomas L. Heath. 2nd ed. 2 vols. New York: Dover.

Forrest, Peter. 2008. Relativity, the Passage of Time and the Cosmic Clock. In Dieks 2008, 245–253.

Friedman, Michael. 1983. *Foundations of Spacetime Theories*. Princeton: Princeton University Press.

Galileo Galilei. 1989 [1638]. *Two New Sciences*. Translated, with a New Introduction and Notes, by Stillman Drake. 2nd ed. Toronto: Wall and Emerson.

Grosholz, Emily R. 2007. *Representation and Productive Ambiguity in Mathematics and the Sciences*. Oxford: Oxford University.

Guicciardini, Niccolò. 1999. *Reading the Principia*. Cambridge: Cambridge University Press.

———. 2002. Analysis and Synthesis in Newton's Mathematical Work. In *The Cambridge Companion to Newton*, ed. I. Bernard Cohen and George Smith, 308–328. Cambridge: Cambridge University Press.

———. 2004. Geometry and Mechanics in the Preface to Newton's *Principia*: A Criticism of Descartes' *Géométrie*. *Graduate Faculty Philosophy Journal* 25 (2): 119–159.

Gutfreund, Hanoch, and Jürgen Renn. 2015. *The Road to Relativity: The History and Meaning of Einstein's*, The Foundation of General Relativity. Princeton: Princeton University Press.

Hartle, James. 2003. *Gravity: An Introduction to Einstein's General Relativity*. San Francisco: Addison Wesley.

Heath, Thomas L. 2009 [1910]. *Diophantus of Alexandria: A Study in the History of Algebra*. 2nd ed. Mansfield: Martino Publishing.

Heelan, Patrick. 1987. Husserl's Later Philosophy of Natural Science. *Philosophy of Science* 54: 368–390.

Hopkins, Burt. 2011. *The Origin of the Logic of Symbolic Mathematics: Edmund Husserl and Jacob Klein*. Bloomington: Indiana University Press.

Howard, Don, and John Stachel, eds. 1989. *Einstein and the History of General Relativity*. Boston: Birkhäuser.

Husserl, Edmund. 1970 [1939]. The Origin of Geometry. In *The Crisis of the European Sciences and Transcendental Phenomenology*, translated with an Introduction by David Carr, Appendix VI, 353–378. Evanston: Northwestern University Press.

———. 1978 [1929]. *Formal and Transcendental Logic*. Translated by Dorion Cairns. The Hague: Martinus Nijhoff.

Janssen, Michel. 2009. Drawing the Line Between Kinematics and Dynamics in Special Relativity. *Studies in History and Philosophy of Modern Physics* 40: 26–52. https://doi.org/10.1016/j.shpsb.2008.06.004.

Kinsman, Thomas, Mark Fairchild, and Jeff Pelz. 2012. *Color Is Not a Metric Space*, Multidisciplinary Vision Research Labs. Rochester: Rochester Institute of Technology. http://rit-mcsl.org/fairchild/PDFs/PRO38.pdf. Accessed 22 July 2017.

Klein, Jacob. 1985. *Lectures and Essays*, edited by Robert W. Williamson and Elliott Zuckerman. Annapolis: St. John's College Press.

———. 1992 [1934–1936]. *Greek Mathematical Thought and the Origin of Algebra*. Translated by Eva Brann. New York: Dover. Republication of original edition, Cambridge, MA: MIT Press, 1968.

Kline, Morris. 1972. *Mathematical Thought from Ancient to Modern Times*. New York: Oxford University Press.

Lehmkuhl, Dennis. 2014. Why Einstein Did Not Believe That General Relativity Geometrizes Gravity. *Studies in History and Philosophy of Modern Physics* 46: 316–326. https://doi.org/10.1016/j.shpsb.2013.08.002. Accessed 18 Mar 2014.

Leibniz, G.W. 1989. *Philosophical Essays*. Translated by Roger Ariew and Daniel Garber. Indianapolis: Hackett.

Martínez, Alberto. 2009. *Kinematics: The Lost Origins of Einstein's Relativity.* Baltimore: Johns Hopkins University Press.

Mazzolini, Renato G., ed. 1993. *Non-verbal Communication in Science Prior to 1900.* Firenze: Leo S. Olschki.

Mermin, N. David. 1984. Relativity Without Light. *American Journal of Physics* 52 (119): 119–124. https://doi.org/10.1119/1.13917.

Meyerson, Émile. 1985 [1925]. *The Relativistic Deduction.* Translated by David A. Sipfle and Mary-Alice Sipfle. Dordrecht: D. Reidel.

Minkowski, Hermann. 1952 [1909]. Space and Time. In Einstein et al 1952, 73–96.

———. 2012a. *Space and Time: Minkowski's Papers on Relativity.* Edited by Vesselin Petkov, translated by Fritz Lewertoff and Vesselin Petkov. Montreal: Minkowski Institute Press.

———. 2012b. The Fundamental Equations for Electromagnetic Processes in Moving Bodies. In Minkowski 2012a, 51–110.

———. 2012c. Space and Time. In Minkowski 2012a, 111–125.

Nerlich, Graham. 2013. *Einstein's Genie: Spacetime Out of the Bottle.* Montreal: Minkowski Institute Press.

Newton, Isaac. 1769 [1707]. *Universal Arithmetic.* Translated by Ralphson. Revised and corrected by Cunn. London.

———. 1846 [1726]. *Newton's Principia: The Mathematical Principles of Natural Philosophy.* 3rd ed. First American Edition of Andrew Motte translation of 1729. New York.

———. 1934 [1726]. *Principia.* 3rd ed. 2 vols. Translated by Andrew Motte in 1729. Translations revised with a historical and explanatory appendix by Florian Cajori. Berkeley: University of California Press.

———. 1972 [1726]. *Philosophiae Naturalis Principia Mathematica.* 3rd ed. (1726) with Variant Readings. Assembled and edited by Alexandre Koyré and I. Bernard Cohen with the assistance of Anne Whitman. 2 vols. Cambridge, MA: Harvard University Press.

———. 1999 [1726]. *The Principia: Mathematical Principles of Natural Philosophy.* 3rd ed. Translated by I. Bernard Cohen and Anne Whitman. Berkeley: University of California.

Norton, John. 1987. Einstein, the Hole Argument and the Reality of Space. In *Measurement, Realism and Objectivity*, edited by John Forge, 153–188. Dordrecht: D. Reidel.

———. 1989. What Was Einstein's Principle of Equivalence? In Howard and Stachel 1989, 5–47.

———. 1992. The Physical Content of General Covariance. In *Studies in the History of General Relativity*, edited by Jean Eisenstadt and A.J. Kox, 281–315. Boston: Birkhauser.

———. 1993. General Covariance and the Foundations of General Relativity: Eight Decades of Dispute. *Reports on the Progress of Physics* 56: 791–858.

Pemberton, Henry. 1728. *A View of Sir Isaac Newton's Philosophy*. London: S. Palmer.

Petkov, Vesselin. 2005. *Relativity and the Nature of Spacetime*. Berlin: Springer.

———. 2012. The not-fully-appreciated Minkowski. Introduction to Minkowski 2012a, 1–37.

Putnam, Hilary. 1967. Time and Physical Geometry. *Journal of Philosophy* 64 (8): 240–247.

Pyenson, Lewis. 1977. Hermann Minkowski and Einstein's Special Theory of Relativity. *Archive for History of Exact Sciences* 17 (1) (31.V): 71–95. http://www.jstor.org/stable/41133480.

Roche, John. 1993. The Semantics of Graphics in Mathematical Natural Philosophy. In *Non-verbal Communication in Science Prior to 1900*, edited by R.G. Mazzolini, 197–233. Firenze: Leo S. Olschki.

———. 1998. *The Mathematics of Measurement*. London: Athlone Press.

Ryckman, Thomas. 2005. *The Reign of Relativity: Philosophy in Physics 1915–1925*. Oxford: Oxford University Press.

Rynasiewicz, Robert. 2012. Simultaneity, Convention, and Gauge Freedom. *Studies in History and Philosophy of Modern Physics* 43: 90–94. https://doi.org/10.1016/j.shpsb.2012.01.004.

Savitt, Steven. 2011. Time in the Special Theory of Relativity. In *The Oxford Handbook of Philosophy of Time*, edited by Craig Callender, 546–570. Oxford: Oxford University Press.

Schutz, Bernard F. 1985. *A First Course in General Relativity*. Cambridge: Cambridge University Press.

Sommerfeld, Arnold. 1970 [1949]. To Albert Einstein's Seventieth Birthday. In *Albert Einstein: Philosopher-Scientist*, edited by Paul Arthur Schilpp, 99–105. LaSalle: Open Court.

Stachel, John. 1989a. The Rigidly Rotating Disk as the 'Missing Link' in the History of General Relativity. In Howard and Stachel 1989, 48–62. Boston: Birkhäuser.

———. 1989b. Einstein's Search for General Covariance, 1912–1915. In Howard and Stachel 1989, 63–100.

Stallo, J.B. 1960 [1881]. *The Concepts and Theories of Modern Physics*. 3rd ed. Edited by Percy W. Bridgman. Cambridge, MA: Harvard University Press.

Stevin, Simon. 1585. *Arithmetique*. Leiden.

Synge, J.L. 1960. *Relativity: The General Theory*. Amsterdam: North-Holland Publishing Company.

Torretti, Roberto. 1983. *Relativity and Geometry*. Oxford: Pergamon.

Unger, Roberto Mangabeira, and Lee Smolin. 2015. *The Singular Universe and the Reality of Time*. Cambridge: Cambridge University Press.

Van Flandern, Tom. 2008. Global Positioning System and the Twins' Paradox. In Craig and Smith 2008, 212–228.

Vieta, François. 1992 [1646]. *Introduction to the Analytical Art*. Translated by J. Winfree Smith. In Klein 1968, 315–353.

———. 2006 [1646]. *The Analytical Art*. Translated by T. Richard Witmer. Mineola: Dover. Unabridged republication of the work originally published by The Kent State University Press. Kent, 1983.

Vishwakarma, Ram Gopal. 2012. On the Relativistic Formulation of Matter. *Astrophysics and Space Science* 340: 373–379. https://doi.org/10.1007/s105509-012-1051-4.

Walter, Scott. 1999. Minkowski, Mathematicians, and the Mathematical Theory of Relativity. In *The Expanding Worlds of General Relativity* (Einstein Studies Volume 7), edited by Hubert Goenner, Jürgen Renn, Jim Ritter, and Tilman Sauer, 45–86. Boston: Birkhäuser.

———. 2014. The Historical Origins of Spacetime. In *Springer Handbook of Spacetime*, edited by A. Ashtekar and V. Petkov, 27–38. Berlin: Springer.

Weyl, Hermann. 1922. *Space, Time, Matter*. Translated from the German by Henry L. Brose. New York: Dover.

———. 1949. *Philosophy of Mathematics and Natural Science*. Princeton: Princeton University Press.

Will, Clifford. 1993. *Theory and Experiment in Gravitational Physics*. Rev. ed. Cambridge: Cambridge University Press.

Winnie, John A. 1970. Special Relativity Without One-Way Velocity Assumptions: Part I. *Philosophy of Science* 37 (1): 81–99. http://www.jstor.org/stable/186029.

Index[1]

[1] Note: Page numbers followed by 'n' refer to notes.

J. K. Cosgrove, *Relativity without Spacetime*,
https://doi.org/10.1007/978-3-319-72631-1

The manufacturer's authorised representative in the EU is Springer Nature Customer Service Centre GmbH, Europaplatz 3, 69115 Heidelberg, Germany. If you have any concerns regarding our products, please contact ProductSafety@springernature.com

Printed and bound by CPI Group (UK) Ltd, Croydon, CR0 4YY
05/10/2026
02228128-0001